THE
COMPLETE
IDIOT'S
GUIDE® TO

Theories of the Universe

by Gary F. Moring

ALPHA

A member of Penguin Group (USA) Inc.

To Fritjof Capra, Gary Zukov, and Ken Wilbur for starting me on the path 20 years ago that led to the profound wisdom inherent in the universe through the blending of science, the Eastern traditions, and the role of human consciousness.

International Standard Book Number: 0-02-864242-2
Library of Congress Catalog Card Number: 2001097251

07 06 05   8 7 6 5

Interpretation of the printing code: The rightmost number of the first series of numbers is the year of the book's printing; the rightmost number of the second series of numbers is the number of the book's printing. For example, a printing code of 02-1 shows that the first printing occurred in 2002.

Printed in the United States of America

Note: This publication contains the opinions and ideas of its author. It is intended to provide helpful and informative material on the subject matter covered. It is sold with the understanding that the author and publisher are not engaged in rendering professional services in the book. If the reader requires personal assistance or advice, a competent professional should be consulted.

The author and publisher specifically disclaim any responsibility for any liability, loss, or risk, personal or otherwise, which is incurred as a consequence, directly or indirectly, of the use and application of any of the contents of this book.

Publisher
Marie Butler-Knight

Product Manager
Phil Kitchel

Managing Editor
Jennifer Chisholm

Acquisitions Editor
Mike Sanders

Development Editor
Tom Stevens

Production Editor
Billy Fields

Copy Editor
Cari Luna

Illustrator
Jody Schaeffer

Cover Designers
Mike Freeland
Kevin Spear

Book Designers
Scott Cook and Amy Adams of DesignLab

Indexer
Angie Bess

Layout/Proofreading
Svetlana Dominguez
Rebecca Harmon

Contents at a Glance

Appendixes

Contents

Introduction

Welcome to the exciting world of cosmetology. In the pages of this book you'll find out how to cut the latest hairstyles, apply make-up and oops ... wrong -ology! Let's start again. Actually, when I told a few of my friends that I was writing a book about cosmology, some of them were amazed to think that I had gone to a beautician school. What this little story illustrates, is that sometimes we may think we know what we know to be true, but can later find out that we were mistaken. In other words, just because you believe something is true, doesn't mean it really is. This is especially true when we seek to understand how our universe began, evolved, and how it operates.

The study of cosmology can be approached from a number of perspectives. Generally, we associate it with some branch of science, such as astronomy, physics, astrophysics, or quantum theory. Just as often it can also be found in the realm of philosophy or religion. And by looking at it from a historical point of view, you can see how it has changed not only over time, but also from culture to culture.

One of the goals of this book is to provide you with a good overview of the latest cosmological theories in science. I will also give you a look at the history of cosmology from cultural, literary, religious, and philosophical perspectives. In such a broad topic as this, an interdisciplinary or multifaceted approach works best. It will offer you the clearest access to gaining a good understanding of how deep the current of cosmology runs in our everyday world. Here's an outline of what will be covered in each section:

Part 1, "In the Beginning ...," looks at the earliest creation stories from around the world. We will trace the development of the first theories of the origin of the universe up through the birth of the scientific method in the sixteenth century.

Part 2, "Science Comes of Age, but Isn't 21 Yet," discusses the birth of the scientific revolution, the influence of the alchemical tradition on Newton and Boyle, and provides you with a background to some of the areas in physics that relate to cosmology.

Part 3, "And in This Corner ...," examines some of the problems that classical physics can't explain, delves into the debate between evolution theory and creationism, looks at some unusual characteristics of light, and introduces you to Einstein's theory of special relativity.

Part 4, "From Here to Infinity," acquaints you with the strange world of quantum mechanics. You'll meet the Standard model of the microcosm and macrocosm and examine some of Einstein's other theories.

Part 5, "Supersymmetry, Superstrings, and Holograms," brings home the latest theories that are the best candidates for the ultimate theory. The holographic paradigm will also bring the role of human consciousness even more into the equation.

Part 6, "Old Endings and New Beginnings," completes our look at theories of the universe by examining how the East and the West see how the universe will end. Along the way we'll take an in-depth look at symbols and their potential to unify a wide range of theories.

To help you along the way, you'll find the following four sidebars:

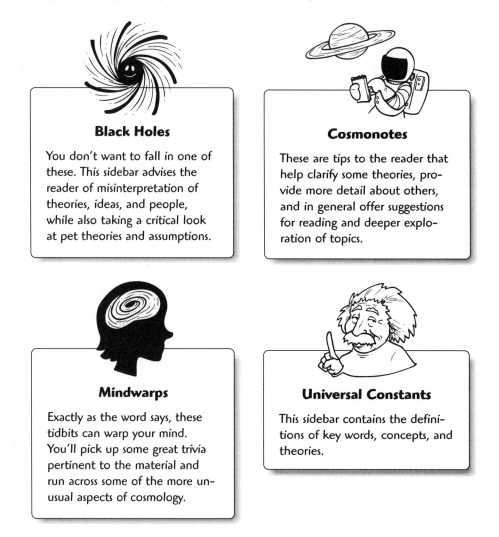

Black Holes

You don't want to fall in one of these. This sidebar advises the reader of misinterpretation of theories, ideas, and people, while also taking a critical look at pet theories and assumptions.

Cosmonotes

These are tips to the reader that help clarify some theories, provide more detail about others, and in general offer suggestions for reading and deeper exploration of topics.

Mindwarps

Exactly as the word says, these tidbits can warp your mind. You'll pick up some great trivia pertinent to the material and run across some of the more unusual aspects of cosmology.

Universal Constants

This sidebar contains the definitions of key words, concepts, and theories.

Acknowledgments

Although there have been many influential people that have inspired me in this subject, this book could not have been made without the support of the following individuals:

Andre Abecassis, my agent, who talked me into another book and kept me from abandoning this project; Larry Abrahamsen, for creatively putting the figures on his computer; Daniel "Bugso" Wajerski who sent me the inspiring cartoon; Alice Bailey and D.K. for the esoteric material that sparked this book; Katie Carrin for her great bodywork sessions, Sam Mills who supported me in more ways than one, and Mike Hemingway who was always there when I needed a break; Jennifer, Larry, Katie, and my parents, Frank and Betty Moring, for taking me out to lunch on Kathy's week off; and, last but not least, my wife, Kathy, who not only supported me in this endeavor, but contributed to it as well.

Trademarks

All terms mentioned in this book that are known to be or are suspected of being trademarks or service marks have been appropriately capitalized. Alpha Books and Penguin Group (USA), Inc.., cannot attest to the accuracy of this information. Use of a term in this book should not be regarded as affecting the validity of any trademark or service mark.

Part 1
In the Beginning ...

Science estimates the age of our universe to be between 12 and 15 billion years old. That's quite a long time, considering we have only been around for a fraction of that entire period. But from the very beginning, when our ancient ancestors looked up into the night sky, they wondered, "How did it all start?" Is there some purpose for us being here? Are we the only people in this vast cosmos, or are there other beings out there? Perhaps there are gods and goddesses, or maybe other people like us.

We have come up with some remarkable ideas and theories to answer these age-old questions. Religion has sought to provide us with some of these answers. In the first part of the book, I'll spend some time looking at the early creation stories from various cultural traditions. After that, I'll show you the transitions that began to take place as rational thought started to influence our understanding. I think you'll find an interesting correlation among some of the creation stories. Our cultural roots may have a lot more similarity than we usually think.

Once Upon a Time

There are two ways in which we have tried to understand the origin and structure of the universe: The first method is through the application of numerical symbols, such as numbers; the second, by studying the stars. These two systems of thought have their beginnings in our earliest civilizations and have stayed with us to the present day, although now they are much more sophisticated and complex. It's important to see how these two methods began so that you can trace the evolution of not only their development but our understanding of the universe as well.

We will begin our search for understanding by exploring some of the civilizations of the ancient Near East. Those are the cultures of Mesopotamia and Judaism. These people and their interpretations of the universe have laid the foundation for many of the cultural beliefs we still hold today.

Cosmology, Cosmogony, and the Cosmos

Before we begin discussing these early civilizations, an important point needs to be made. The word *cosmos* comes from the Greek word *kosmos*, meaning universe, world, and harmony. The study of the origin of the universe is called cosmogony. The latter part of the word, *gony*, comes from the Greek word, *gonos*, meaning to produce. If you put it all together, *kosmos*, universe + *gonos*, to produce, you get a universal produce market. No, just kidding. What we really get is the generation or the producing of the universe.

So *cosmogony* is the study of the origin of the universe. Then what is *cosmology?* That pertains to the study of the evolution and structure of the universe. This may seem like splitting hairs to you but it's good to know the difference between the two. Some people use cosmology to encompass the study of all the aspects of the universe and don't differentiate that much between the two. It might be easier for us too, so from now on I'll just use cosmology to refer to the study of any aspect of the universe.

As I mentioned earlier, there are two main systems of thought that have been used to provide structure, order, and meaning to the universe—numerical and astronomical. By numerical I mean the application of numbers to give structure and order, not the use of mathematics to explain a process or define natural laws—those ideas came much later. As you will soon see, numbers formed the underlying structure that gave us the particular order and rhythm by which we live our lives.

Universal Constants

The combining form, **cosmo,** has slightly different meanings, depending on how it's used as the root in a word. For example, in words like cosmology, cosmogony, and cosmography it means "universe." In other words like cosmopolitan, cosmopolite it means "worldly." And the word cosmos by itself doesn't just mean "universe," but encompasses the notion of it being harmonious and orderly.

The second system of thought, the astronomical system, was developed in direct correlation with the use of numbers, but it also went beyond that. While numbers gave structure and order, the observation and study of the heavens gave meaning and placed us within the overall scheme of the universe. We will explore that method in the next chapter. For now, let's take a look at how numbers were first utilized and how this all unfolded.

Mesopotamia, or Where It All Began

Civilization is considered to have started in an area known as the Fertile Crescent. Cradled by the Tigris and Euphrates Rivers in what is today Iraq, the Sumerians of Mesopotamia established the earliest known society in which people could read and write. Mesopotamia is a Greek word meaning the land between two rivers. As far back as 10000 B.C.E. there was a great arc of trade, a huge travel route from Sumer through Syria and Canaan into Egypt. Cities such as Jericho, built around 8500 B.C.E., were centers of trade, and many of the first cities were established along this route.

The Sumerians came from the east or from the mountains of Elam perhaps as early as 5000 B.C.E., to the swamps at the head of the Persian Gulf. They drained the swamps, developed flood control, invented the wheel, and established permanent agriculture. As successful trading continued to develop with the surrounding areas, Sumerian villages and settlements grew into prosperous city-states, such as Ur, Eridu, Lagash, Nippur, and Uruk.

It's ironic, or perhaps inevitable, that the first words that people ever wrote were entries in a ledger; lists of goods stored in a Mesopotamian temple recorded in pictographic abstract markings, called cuneiform, impressed onto shiny pillows of clay with a wedge-shaped stick.

So You Want to Be a Hero!

The oldest written story also comes from Mesopotamia. You may have heard of it, "The Epic of Gilgamesh." It predates any other ancient myths, including the stories in the Old Testament. It tells the story of Gilgamesh, a historical king of Uruk in Babylonia, where he lived around 2700 B.C.E. The fullest surviving version, told on 12 clay cuneiform tablets, was found in the ruins of the library at Ashurbanipal, in Nineveh.

In the myth, Gilgamesh is presented in superhuman form, a powerful being who had all knowledge and built the great city of Uruk. Each tablet relates one of his many heroic adventures and it is in one of these that we find the first story of the flood, the same flood that's related in the Book of Genesis. And this brings us to the first method I've previously talked about, the usage of numbers to begin providing structure to our understanding of the universe. As you will see shortly, the number 7 will play a fundamental role in how it's all organized.

Cosmonotes

When we speak about stories from ancient cultures we often use the word myth. The meaning of the word is not to be confused with the idea of a falsehood or untrue statement. It refers to the sacred stories from any culture that contain elements that later came to form the core of religious beliefs and traditions.

In tablet 11, the tale is told of a dream that another king had. In this dream he is told to build a huge boat, because the gods plan on destroying the earth in a great flood. The king, called Utnapishtim (say that 10 times fast), loads the boat with gold, silver and every living thing. The flood lasts for seven days and seven nights, with light returning to earth after seven days. The huge boat comes to rest on the top of Mount Nimush, and remains there for seven days. On the seventh day all the living creatures are released.

In this brief recounting of the flood story you can see how one specific number is beginning to set up a pattern of similarity. Let's take a look at the Mesopotamia creation story to see how this number lends an even deeper structure to the whole thing.

Mindwarps

The story of the flood is a well-known one. Many of us are familiar with it from the Book of Genesis. But that is not the only source in which it can be found. You can find it in the *Popul Vuh,* the ancient sacred book of the Maya. Plato also gives an account of it in his dialogue, *Critias.* And as you've already read, it is also found in the Mesopotamian myth, "The Epic of Gilgamesh." As a matter of fact, it's found in many stories from cultures all around the world.

Family Feud

The *Enuma Elish,* the story of the creation of Mesopotamia and Man, was recited in the temples of Ur and Haran as in every Mesopotamian temple every year for some 4,000 years and more. This story is also ordered in seven units, in seven generations of a family of gods. What is interesting is how these seven generations of gods, relate through correspondence to the seven days of creation found in the Book of Genesis, which will be coming up shortly.

The creation of Mesopotamia is related through the tale of seven generations of family passions. It begins with Apsu, the great patriarch who resolves to kill all of his descendants for making too much noise. But Ea, his great-grandson kills him first, rousing Tiamat, Apsu's wife, to fury. Leading a host of demons, she and a new husband declare war upon the rest of the family. Marduk, the youngest of them all, offers to be their champion if they will accept him as their king. So Marduk goes off to war, routs the demon army and kills the mother goddess. He splits her carcass in half to make the heaven and earth, and, from the arch of her legs, a vault to support the sky. He buries her head under a mountain, piercing her eyes so that the Tigris and Euphrates can flow from them.

There's much more to the whole story, but I wanted to spare you the gory details. That's a pretty horrible account compared to the majestic telling of creation we find in Genesis and some other cultural stories. But many ancient civilizations related their myths in terms of family dynamics. It's not unusual to find tales of incest, murder, and other family feuding taking place. You'll find examples of that in all sacred texts, from the stories in the Old Testament to the *Popul Vuh* of the Maya. How better to relate important chronicles than through a medium everyone is familiar with, the family?

The Magnificent Seven

Whenever I think of the Magnificent Seven I can't help but hear the old Marlboro cigarette commercial theme song playing in my head. But that's a different magnificent seven. What I'll be exploring with you in this section is the version of creation that many people of Western culture are most familiar with.

Genesis

> In the beginning God created the heaven and the earth. And the earth was without form, and void; and darkness was upon the face of the deep. And the spirit of God moved upon the face of the waters. And God said, Let there be light: and there was light. And God saw the light, that it was good: and God divided the light from the darkness. And God called the light Day and the darkness he called Night. And the evening and the morning were the first day.
>
> —Genesis 1:1–5

God's grand division of light and darkness marks out all the later events of his creation. God has made himself a clock and with it he will measure out the making of his world in seven units—seven days. It breaks down basically like this:

➤ First day, creation of light

➤ Second day, creation of heavens and water

➤ Third day, creation of land and vegetation

➤ Fourth day, creation of bodies of light

➤ Fifth day, creation of creatures of heaven and waters

➤ Sixth day, creation of life on land and its vegetable food, creation of humankind

➤ Seventh day, God rests from all the work

In Judaism, tradition has it that Moses wrote this version of the creation story along with the first five books of the Bible. These books also comprise the *Torah*, the sacred scroll made of parchment that is found in the Tabernacle of any Jewish temple.

Black Holes

It's easy for us here in the West to assume that the beliefs by which we live our lives are the truth. Tradition and years of religious intolerance can blind us to the fact that other traditions can be equally valid as our own. When you study comparative religion and mythology you can sometimes uncover information that can bring into question cherished beliefs. And as I mentioned in the introduction, just because you believe something is true doesn't mean it really is.

Universal Constants

The word **Torah** in Hebrew means "the Teaching." It encompasses the whole body of Jewish religious literature. But it also refers to just the first five books of the Old Testament written on the sacred scrolls of parchment. You may have also heard it referred to as the *Pentateuch*, which is the Greek word for "five books."

Ancient Hebrew Cosmology

The preceding description of the world doesn't share the same scientific view that we have, in which the Earth is one planet around one sun in a universe full of suns and planets. The ancient picture of the universe portrays a world in which the Earth is a disc surrounded by water not only on the sides, but underneath and above as well. A firm bowl (the firmament) keeps the upper waters back but has gates to let the rain and snow through. The Sun, Moon, and stars move in fixed tracks along the underside of this bowl. From below the disc, the waters break through as wells, rivers and the ocean, but the Earth stands firm on pillars sunk into the waters like the pilings of a pier. Deep below the Earth is Sheol, the abode of the dead, which can be entered only through the grave.

In this description of creation the Israelites were no different from other ancient people around them. All ancient cultures placed the earth in the most significant position in relation to the rest of the universe. And I'm sure you can see how the number 7 seems to be a magical number that is used to describe many events. We also find reference to the number 7 in the New Testament as well. There were the last seven words of Jesus on the cross and the Seven Seals of the Book of Revelations, which would be opened at the blast of seven trumpets. But as you'll see in a moment there is more to it than just that.

The Chicken or the Egg

So, how can we relate the Mesopotamian and Hebrew creation stories? Which one came first? They are both similar in how they use the number seven to lay the basis for order and structure. We need to look at the impact that dominant cultures have on surrounding cultures to determine how this all came about.

One way of showing you how this process operates is to use an example of another culture. In Asia, for example, China was the dominant culture for over 2,000 years. Its impact on other countries such as Japan, Korea, Vietnam, and many other smaller countries was reflected in their religion, philosophy, architecture, laws, and especially writing. This process is called *syncretism*.

Remember that the first form of writing, cuneiform, arose in Mesopotamia, and it spread quickly through the merchant cities of the near east, right through the Fertile Crescent down into Egypt. With it went many of the sophisticated ideas of these southern cities. For as well as recording accounts, receipts, and contracts, these little tablets also carried with them notions of civil law, the system and order of Mesopotamian science and literature, a scale of time in years and minutes, and numbering systems in units of 60, which provided a 360-degree circle and the astrological system of the 12 divisions of the zodiac. The Mesopotamian gods also traveled with these tablets, their names and stories, rites and rituals permeated the East.

The Bible tells us that Abraham, the founding father of the Hebrews, traveled from Ur of the *Chaldees* to a city called Haran, and Haran, the cuneiform tablets tell us, stood close by ancient Urfa. With this information it's easy to see that while the Mesopotamian culture was already flourishing, the roots of Judaism were just beginning.

Now let's look at how the process of syncretism occurred between the early Hebrews and Mesopotamia. Just as in the first three days of Genesis'

Universal Constants

The word **syncretism** means the combination or absorption of differing beliefs or practices that are found in religion, philosophy, and other customs. It is a dynamic process that has gone on since the first civilizations and continues to the present day. There is no religion or philosophy that is pure or untouched. All of them have been influenced by outside factors.

Creation, Jehovah makes the elements of the world, so in the *Enuma Elish,* the first three generations of gods are the gods of water, silt, and sky, the elements of the Mesopotamian world. And just as in the fourth, fifth, and sixth days of Genesis' Creation, God makes the animate world, so in the *Enuma Elish,* the fourth, fifth, and sixth generation of gods are the gods of moving things. For example, on the fourth day Jehovah makes the Sun, Moon, and stars and the male deity of the *Enuma Elish* is Anu, the god of the heavens. And finally, just as God rests after these six days and then makes man, so the god Marduk of the sixth generation of *Enuma Elish* creates man so that the Mesopotamian gods might rest.

The two stories, the two universes, are built of the same bricks. They both build the same environment. But while the priestly authors of Genesis obviously knew the Babylonian story, or one very similar, and used its outline, they disagreed with the theology. And that's where this story ends. My intention was to show you how number has gone into providing the basic order and structure of our universe; theology I'll save for another time.

Mindwarps

Chaldea was a province of Babylonia, and the term first made its way into the Bible when it was translated into Greek, because it's actually the Greek name for that area. It's a term that you will come across later in this book, too, because it's associated with astrology, the forerunner of astronomy.

Compass Headings and Complementary Opposites

I've spent a good portion of this first chapter relating the cosmology of just two ancient cultures because these two have had the most impact on how the West has developed and understood some of the most basic aspects of how the universe began, at least from a religious point of view. For the remainder of this chapter let me paint you a picture of how some other cultures have used numbers to lay out their universe.

All cosmologies, or at least all early conceptions of the universe, have a religious aspect. Native American, Chinese, and Egyptian cosmologies are no exception, and each expressed their versions in various ways. Most included some numerical symbolism similar to what we've already discussed, while others relied on purely *anthropomorphic* tales of animal or family interactions.

Universal Constants

The word **anthropomorphic** derived from the verb "to anthropomorphize," which means "to attribute humanlike shape or characteristics to a god, goddess, or animal." It is a common method we use to give form to transcendent concepts so that we can relate to them.

The Four Directions

The oldest and most widespread numerical theme among Native Americans, the great indigenous tribal cultures of North and South America, is the four directions of space. Myths and rituals from a wide range of these tribal cultures, extending from Canada to Peru, incorporate the number 4, its multiples, and the 4 directions. Paintings in caves and on rocks from a similar range of cultures document that the four directions were crucial to having a visual understanding of the cosmos for many tribal people. Here's an example of just how important this number was—the excerpt is from a discussion by a Lakota wise man recorded at the turn of the century:

> In former times the Lakota grouped everything by fours. This was because they recognized the four directions: the west, the north, the east, and the south; four divisions of time: the day, the night, the moon and the year; four parts in everything that grows from the ground: the roots, the stem, the leaves, and the fruit; four kinds of things that breath: those that crawl, those that fly, those that walk on four legs, and those that walk on two legs; four things above the world: the Sun, the Moon, the sky, and the stars; four kinds of gods: the great, the associates of the great, the gods below them, and the spiritkind; four periods of human life: babyhood, childhood, adulthood, and old age. And finally, mankind has four fingers on each hand, four toes on each foot, and the thumbs and great toes taken together all form four. Since the Great Spirit caused everything to be in fours, mankind should do everything possible in fours.

You can see by reading this that the four directions came first in this Lakota discussion of fourness, which reinforces the idea that the four spatial directions (north, south, east, and west) are primary in this theme.

Mindwarps

For the Hopi and other Native American peoples, the four directions are not the familiar cardinal directions of the compass—north, south, east, and west—but are directly related to the observation of the rising and setting sun at solstices. In other words, the four directions refer to the place of sunset at the summer solstice, the place of sunset at winter solstice, the place of sunrise at winter solstice, and the place of sunrise at the summer solstice.

The theme of a primordial fourness of space is not limited to North American Native Americans. Four is also used in the cosmologies of the Aztec and Maya peoples of Mesoamerica, and is very similar to the Lakota tradition.

Heaven and Earth

In Chinese cosmology, the earliest record that describes the origin of the universe dates from the period of the legendary Yellow Emperor, 2698–2359 B.C.E. It was more of a philosophical system of explanation rather than a strictly religious one, although the basic concepts later became part of Taoism and Confucianism. It begins by saying that the universe was originally chaos, and after a period of time broke into two dynamic forces, yin and yang. This duality underlies the entire structure and orders the workings of the universe in a unique way. There is no defined deity or god behind this interactive process; it is simply the fundamental system under which the whole universe operates.

The Chinese yin/yang symbol of complementary opposites.

I'm sure that you are familiar with the symbol shown in the picture. It's common to see it on T-shirts, jewelry, tattoos, and is often associated with martial arts. But few people understand the deep significance it has. To begin with, on a simple level, it is a system of two interacting, complementary energies. The following table shows some of its associations.

Yin and Yang Complements

Yin	Yang
Earth	Heaven
Female	Male
Dark	Light
Soft	Hard
Wet	Dry

Yin	Yang
Down	Up
Left	Right
Exhale	Inhale
Negative	Positive

That's probably enough to give you the general idea. I'm sure you could come up with many more of your own. The important thing to realize is that these are not fixed energies, but are in constant motion and don't really define themselves as black and white as it may seem. The symbol itself contains seeds of the other within it, so neither are completely one or the other. Each is also needed to define the other one, because without one you don't have the other. That is one reason why they are considered complementary and not opposites.

If you look at the yin and yang symbol, you will notice that when one color has grown to its largest expression, the other color begins. This is a dynamic rotating cycle. Think of how you can apply this to a 24-hour day, the phases of the Moon, and the seasons of the year.

This concept of yin and yang is not only the basis for their cosmology, but underlies all of Chinese culture. It is applied to traditional medicine, cooking, painting, architecture, and martial arts, just to name a few, and helps to define any and all relations in the human world. It's also the basis for the binary number system that lies at the core of every computer in the world. Whew, it's amazing what just two numbers can do.

We've now had a chance to see how only three numbers—7, 4, and 2—have given structure and order to our universe. There are other numbers we didn't get to, such as 3 and 12. How many associations can you make with those two numbers as well as any others? In the next chapter I will show you the other system I mentioned, astronomical, and how that can be combined with numbers to add another dimension of understanding to cosmology.

Cosmonotes

If you would like to see how these two forces interact in your personal life, I think you would enjoy learning about the *I Ching* (pronounced *yee-jing*), or *Book of Changes*. It is the ancient classic book of wisdom that teaches you how to recognize the unconscious patterns in your life and to become consciously aware of them.

The Least You Need to Know

➤ Cosmology is the study of the evolution and structure of the universe, while cosmogony deals specifically with the study of its origin.

➤ All cosmologies of early civilizations were religious in nature.

➤ The rhythm and order by which we live our daily lives has its roots in the cosmology of ancient Mesopotamia.

➤ The use of numbers was the first method ancient cultures applied to organize their world.

➤ Cuneiform was the first system of writing. It was developed in Mesopotamia around 3500 B.C.E.

The Music of the Spheres

> ### In This Chapter
>
> ➤ The sacred numbers
>
> ➤ The origin of astrology
>
> ➤ Pythagorean music
>
> ➤ Plato's cosmology

We now take a look at the contributions made by some of the early Greek thinkers. Many aspects of Western civilization began with their ideas. Our notions of politics, philosophy, science, medicine, psychology, art, and astronomy all can be traced back to this culture.

While there are a number of these thinkers we could discuss, I plan on focusing on just two in this chapter, Pythagoras and Plato. Since we're looking at theories of cosmology, these two philosophers more than any others used numbers in conjunction with observations made about the heavens to explain the workings of the universe.

In the last chapter you saw how numbers were first used by ancient cultures to provide a sense of order to their world. We now move forward in time to about 550 B.C.E. and take a close look at Pythagoras, mathematician extraordinaire.

The Essence of All Things

The following is a quote from one of Aristotle's works in which he sums up the Pythagorean attitude toward numbers:

> The Pythagoreans, as they were called, devoted themselves to mathematics; they were the first to take up this study, and having been brought up in it they thought that its principles were the principles of all things. Since, of these principles, numbers are by nature the first, and in numbers they seemed to see many resemblances to things that exist; more than just air, earth, fire, and water, but such things as justice, soul, reason and opportunity.

As you can see, numbers for them were much more than just a way to organize and order the universe. They were completely and absolutely in love with them. The Pythagoreans had a uniquely different approach to understanding numbers. For us today they are just symbols or signs, used to describe a specific quantity or amount. But for them, a number was a living, universal principle whose secrets could only be revealed by careful study. In a certain sense they were considered sacred.

Mindwarps

Around 600 B.C.E., a major transition began taking place in Greece. The philosophers of this time started to seek a deeper understanding of the origin and operation of the world. For them the ancient myths just didn't answer the questions they were asking. They thought it was important to penetrate beneath the appearances of nature, to find her inner workings. What composed all of the various things they saw around them? Were they made up of the four basic elements—earth, air, fire, and water? Some thought so. Others came up with the idea of very tiny, indivisible particles, that they called atoms. These individuals were the closest thing to our scientists of today.

As a young man Pythagoras (c. 582–507 B.C.E.) traveled and studied in the temples of Phoenicia, Egypt, and Babylonia. He spent 22 years in Egypt acquiring as much wisdom as possible from the priests and was trained in geometry and astronomy. In Babylonia, where he studied the teachings of the Magi for 12 years, he learned all he could about music, mathematics, astrology, and all other sciences. He finally returned to his birthplace, the island of Samos, and eventually settled in southern Italy where he lectured on philosophy and mathematics. He started an academy that gradually formed into a brotherhood called the *Order of the Pythagoreans.*

Numbers with Feelings

Did you ever think that numbers could be considered masculine or feminine? Well the Pythagoreans did! Odd numbers were considered masculine and even numbers feminine. And there's more to it than that. Here is a breakdown of just what they considered the first 10 numbers to be.

To begin with, the Pythagoreans didn't consider 1 as a number at all, but as a principle, a unity, underlying everything. It was the source of all things, too. Here is a description of it from one of Pythagoras's students, Theon of Smyna:

> Unity is the principle of all things and the most dominant of all that is: all things emanate from it and it emanates from nothing. It is indivisible and it is everything in power. It is immutable and never departs from its own nature through multiplication ($1 \times 1 = 1$). Everything that is intelligible and not yet created exists in it; the nature of ideas, God himself, the soul, the beautiful and the good, and every intelligible essence, such as beauty itself, justice itself, equality itself, for we conceive each of these as being one and as existing in itself.

Each of the numbers after one has a lengthy description as well, but in the following list, I'll give you the short abbreviated version. They can get pretty complicated.

➤ 1   The monad, a point, is the source of all numbers. It's good, desirable, essential, and indivisible.

➤ 2   The dyad, a line, has diversity. It is the loss of unity—a duality—a number of excess and defect and the first feminine number. (Don't shoot me, ladies. These aren't my definitions.)

➤ **3** The triad, a plane, is composed of unity (1) and diversity (2) and restores them to harmony. It is the first odd, masculine number.

➤ **4** The tetrad, a solid, is the first feminine square (2 × 2), and it represents justice, is steadfast, and a perfect square. It also is the number of the elements, the seasons, the ages of man, lunar phases, and virtues.

➤ **5** The pentad is the masculine marriage number, because it unites the first female number (2) and the first male number (3) by addition. It has the virtue of being incorruptible, because all multiples of 5 end in 5. (I'm not quite sure what that has to do with it being incorruptible, but then, I'm not a Pythagorean.)

➤ **6** The hexad, the feminine marriage number, unifies 2 and 3 through multiplication. It's also considered the first perfect number and is the area of a 3-4-5 triangle.

➤ **7** The heptad is associated with the Greek maiden goddess, Athene. It's considered a virgin number because it has no products, and a circle can never be broken into 7 equal parts. (Hmmm, I'll have to think about that one.)

➤ **8** The octad is the first cube: 2 × 2 × 2 = 8.

➤ **9** The nonad, the first masculine cube, is also considered incorruptible, because however often it's multiplied it reproduces itself. (No comment.)

➤ **10** The decad contains all the numbers, because after it, the numbers just repeat themselves. It's also the number of our fingers and toes.

Mindwarps

Pythagoras, like Socrates, never wrote anything, at least anything that we know of. His teachings were all oral, and what we do know about him came from the writings of his students. He is known to be the first person to call himself a philosopher, a lover of wisdom.

So as you can see, numbers had a wide range of symbolism. Plato, who we will be coming to shortly, advanced this symbolism even more. But for now, let's see how Pythagoras applied this to music, and then to the heavens.

Harmonies of Heaven and Earth

We saw in Chapter 1, "Once Upon a Time," that ancient Chinese cosmologies defined the world in terms of duality, yin and yang. The Pythagoreans did so as well. The following is a list of the ten dualities. But the amazing thing about this list is that a number is associated with many of the qualities.

Limited	Unlimited
Odd	Even

One	Many
Right	Left
Male	Female
Rest	Motion
Straight	Curved
Light	Dark
Good	Bad
Square	Oblong

I think some people nowadays are likely to consider the discussion on the last few pages as poetic, superstitious, and not based in the concrete world of reality. That maybe so, but for the Pythagoreans, *numerology* was a legitimate form of mathematics.

Universal Constants

The study of symbolism of numbers is known as **numerology.** It was not limited in its usage to the Pythagoreans. It can be found in many symbolic systems around the world. It forms the root of the Cabalah, the mystical system of Judaism. It also is the basis for the *I Ching,* the Chinese book of wisdom. And it is intimately related to Western and Eastern astrological systems as well as the Greek and Hebrew alphabet.

Up until now, I've spent most of the time discussing the first method by which ancient cultures ordered the universe, namely by using numbers. At this point I'd like to explain how the study of the stars developed into the early system of *astrology,* and then how these two methods combined to give us an ordered, harmonious, and meaningful universe.

Most ancient cultures applied what they saw in the day and night skies to regulate their calendars for agricultural purposes. From the simple observation of the lunar cycle and the seasonal variation in the sunrise point on the local horizon, to the more elaborate calculation of the solstices and equinoxes, people began to see correlations between events on earth and these cycles.

Universal Constants

Astrology is a method of correlating events on earth with the position of the Sun, Moon, planets, and constellations to explain the significance of these events and to predict future ones. Astrologers believe that earthly events, including people's personality traits and things that occur in their lives, are timed by the cycles of these celestial bodies. What happens in the heavens is reflected here on earth.

Back to Babylon

Some celestial events, such as eclipse and the appearance of comets, were seen as omens. These omens were considered to be direct communications from the divine powers, or in other words, from the gods who lived in the heavens. And as mentioned earlier, while most ancient cultures used the observations they made about celestial events for practical purposes, the system of astrology that developed from these observations came from you know where: Mesopotamia.

Astrology developed out of the planetary omen reading practiced in ancient Mesopotamia from at least 2000 B.C.E. onward. Hundreds of clay tablets, inscribed in cuneiform, have been discovered that record the observation and interpretation of various classes of omens. Here's a translation of one of the tablets:

> There will be an eclipse that is evil for Elam and Aharou, but lucky for you my lord, rest happy. It will be seen without Venus, and to the King, my lord, I say there will be an eclipse.

The Mesopotamians kept detailed records of celestial phenomena and the messages from the gods that these were said to represent. After Mesopotamia was conquered by the Persians in 538 B.C.E., and then by Alexander the Great in 331 B.C.E, astrological theory became more of a secular study, a branch of natural science. Astrologers, known as Chaldeans, traveled throughout the Greek and Roman empires practicing their skills.

The Seven Sacred Planets

By the time Pythagoras was developing his numerical system, the astrologers whom he had studied under in Egypt and Mesopotamia had identified what were considered to be the seven sacred planets of the universe. Sacred because there were gods associated with each one. These gods would become the main characters in Greek and Roman mythology. So Pythagoras adopted the symbolism that went with each one and combined that with his numerical system. The following table shows some of the correspondences between numbers and planets.

Number	Planet	Day	Metal
One	Sun	Sunday	Gold
Two	Moon	Monday	Silver
Three	Mars	Tuesday	Iron
Four	Mercury	Wednesday	Quicksilver
Five	Jupiter	Thursday	Tin
Six	Venus	Friday	Copper
Seven	Saturn	Saturday	Lead

Here then are four associations for each planet. Now I know that today the sun and moon aren't considered planets, but back then they were; nobody knew the difference. I added in the other two correspondences, days of the week and metals, because they will have a significant part in the next chapter. Also notice that here again we find the number seven playing its role.

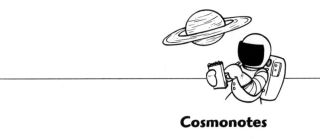

Cosmonotes

Astrology is still very much a part of people's lives today. Just about every newspaper you pick up or magazine you flip through will have a horoscope in it, telling you of portents to come. Although science still considers it quackery, it has made a comeback since the turn of the century. Whether or not science recognizes any validity in it doesn't seem to matter to a lot of people, which is as it should be. Use what works for you.

Music Therapy

A young man from Pythagoras's village had been up all night with friends partying and listening to songs that were of a type well known to incite violence. When the aggravated man saw the girl he loved sneaking back home in the early morning hours from the home of his rival, he decided to burn her house down. Pythagoras happened to be out late that night himself observing the stars, and he walked in on this violent scene. He asked the musicians to change their tune to a more tranquil one, and the young man's madness cooled immediately. He apologized for his behavior and went home in an orderly fashion.

Mindwarps

All of the members of Pythagoras's brotherhood were required to study number in four distinct branches: arithmetic or number itself, geometry or number in space, music or number in time, and astronomy or number in space and time. These four areas of study were called the Quadrivium, which is simply Latin for four roads that meet. The mastery of these four subjects was needed for a Bachelor's degree in the Middle Ages.

This story relates one of the uses that the Pythagoreans had for music. Before they went to bed at night they listened to certain types of music that cleansed them of the noises and restlessness of the day's activities. When they woke up in the morning, they would begin the day by playing songs that dispersed the sluggishness of sleep.

I haven't brought the role of music into the picture yet, because I was saving the best for last. For you see, it's music that finally brings Pythagoras's contributions to a close and introduces you to the music of the spheres.

The Rhythm of the Universe

Aristotle notes:

> The Pythagoreans say that the attributes and ratios of the musical scales were expressible in numbers; since, then, all other things seemed in their whole nature to be modeled after numbers, and numbers seemed to be the first things in the whole of nature, they supposed the elements of numbers to be elements of all things, and the whole heaven to be a musical scale and a number.

In Aristotle's description of what the Pythagoreans thought of music, you can see that for them, music was number and the cosmos was music. They were the first to discover the mathematical relationships between the *harmonic intervals*. Pythagoras did this by experimenting with a *monochord*.

Pythagoras defined three different kinds of music:

➤ Ordinary music made by plucking strings, blowing pipes, essentially all types of music played on instruments.

➤ The continuous but unheard music made by each human being that reflected the harmonious or inharmonious resonance between the soul and the body.

➤ The music made by the cosmos itself, which you already know as the music of the spheres.

The laws of music, defined in mathematical proportions, were very important, for they governed the whole range of the perceptible and the imperceptible universe. Pythagoras's discovery of the musical intervals was not just the beginning of music theory, you could say that it was also the beginning of science. For the first time, through systematic investigation, universal truths could be explained by the use of mathematical symbols.

For those of you who are musically inclined, this is a breakdown of how the music of the spheres worked. Counting outward from the Earth; from the Earth to the Moon was a whole step; from the Moon to Mercury, a half step; Mercury to Venus, another half step; from Venus to the Sun was a minor third, which is equal to three half steps; the Sun to Mars, a whole step; from Jupiter to Saturn a half step; and from Saturn to the sphere of the fixed stars, another minor third. If you played it on a scale it would be C, D, E-flat, E, G, A, B-flat, B, D, otherwise known as the Pythagorean scale. So in other words, the distance between the planets is like a gigantic musical scale. You can look at the keyboard on a piano and play the scale to see what it sounds like.

According to Aristotle, only Pythagoras could hear and comprehend this celestial harmony. The rest of us can't hear it because from the moment of birth the sound fills our ears, and we have become so accustomed to it that we can't distinguish it from

Universal Constants

In music, a **harmonic interval** is the distance between two notes that produces a beautiful sound when the two notes are played in combination. Pythagoras is said to have invented the **monochord.** It consists of a single stretched gut-string with a movable bridge, which enables philosophers and students of music theory to demonstrate the harmonic laws.

Mindwarps

Just in case you didn't know, almost everyone at this time believed that the Earth was at the center of the universe, not the Sun. That's why Pythagoras began counting from the Earth instead of the Sun. It wasn't until the fifteenth century that this belief would be challenged. That discussion will be coming up soon.

other sounds anymore. (Either that or it's because you've listened to music with the headphones turned up too loud.)

Universal Constants

The **Socratic method** or dialectic method is a search for the proper definition of a thing, a definition so clearly described that it can't be refuted. It is a question-and-answer process that goes back and forth between people who are seeking to understand the clearest meaning of the topic in question.

Universal Constants

A **demiurge** is a ruling force or creative power. You can think of it as the same thing as a god. The word also contains the idea of a skilled workman or creator that works for the benefit of people. And for Plato it is the deity that created the material world.

The World Soul

I'm sure you're getting a little tired of all these numbers by now, so let's leave Pythagoras and spend the last few pages of this chapter talking about Plato.

Plato (c. 427–347 B.C.E.) is considered by many philosophers to be one of the greatest and most influential thinkers that have ever lived. According to Alfred North Whitehead, one of the brilliant philosophers of the twentieth century, "All contributions made after Plato are simply footnotes to Philosophy." His overwhelming influence would provide the working outline for Western philosophy for the next 2,000 years.

All of Plato's philosophy can be found in *The Dialogues of Plato*. Socrates plays a major role in many of these and the *Socratic method* becomes the foundation and soul of philosophy.

Plato's cosmology can be found in the dialogue called the *Timaeus*. This dialogue is famous for a number of reasons besides the discussion of the creation of the cosmos and all it contains. In it you can find the first known story of the lost continent of Atlantis.

The First Cause

One of the many questions that people have asked about the origin of the universe is "Why is there a universe at all?" Plato is the first person to answer this question. It would go something like this: To begin with, the world is a decent place, characterized by goodness and order. And such a great place could not have come into being except through some cause, a first cause. Here is where Plato then introduces us to the *demiurge*, the creator of the first cause. This demiurge must have been good and orderly and entirely lacking in envy. And because he was this way, he wanted his creation to be as good and orderly as himself, so he created the world in his own image. Sound familiar?

Now that you have Plato's summary of the origin of the universe, do you think that it's a fair and orderly place? For some of us, probably not, we're a bit more cynical than we used to be. But to understand Plato's view, you have to see the universe in terms of how it's been viewed for the last 2,000 years. It has only been since the Industrial Revolution, or roughly the last 200 years, that this view has slowly disappeared. For most of the history of Western civilization the universe has been seen as a harmonious, ordered whole, perfect and predictable. If you told Plato that you thought the world was a fearful place where random acts of violence took place, he would have thought you had spent too much time eating fast food, going to too many rock concerts, and watching too many soaps on TV.

Cosmonotes

Plato's philosophy strongly influenced the ideas of some of Christianity's greatest theologians. St. Augustine regarded Plato as having prepared him for Christianity and transferred many of Plato's themes to it. If you want to find out more about these early Church fathers, read Augustine's book *Confessions*, as well as works by Anselm and Thomas Aquinas. All three of these men were very strongly inspired by Plato and Aristotle.

You Gotta Have Soul

Within Plato's discussion of the creation of the world he also explains how our soul was created. He states that the first cause is made up of three parts: existence, sameness, and difference. These three parts are mixed in musical ratios, which gives each of us our own unique sound and vibration. He then explains the make-up of the world soul, which the demiurge wrapped around the spherical universe. Plato goes into a rather complicated Pythagorean numerical system to explain that the world soul is explicitly musical. Plato has taken a lot of the ideas of the Pythagoreans to develop a universe that is not only ordered and ultimately musical, but he has also given all of creation an organic, living presence. For him, it is not a universe of dead matter. It is alive, dynamic, and has soul.

You now have a good idea of how Pythagoras and Plato saw the universe. The two methods of numerical symbolism and celestial observation have combined to describe a world that was harmonious, ordered, and alive. In the next chapter I'll introduce you to the world of esoteric philosophy. It is the system of thought that laid the foundation for the advent of the scientific revolution.

The Least You Need to Know

➤ Pythagoras believed that numbers were universal principles, and these principles imbued them with special qualities.

➤ The Pythagoreans combined numerology and astrology to give order, structure, and meaning to the universe.

➤ The astrological system in use today had its beginning in ancient Mesopotamia.

➤ The underlying structure of the universe is defined by mathematical ratios that are musical in nature. It's called the music of the spheres.

➤ Plato's ideas would be the cornerstone of philosophy for over 2,000 years, and would strongly affect the theology of the Church.

Chapter 3

A Brief History of Esoteric Philosophy

In This Chapter

➤ Esoteric vs. exoteric knowledge

➤ The Hermetic philosophy

➤ The Western esoteric tradition

➤ The impact of esoteric thought on the Renaissance mind

What Is Esoteric Philosophy?

Any discussion of the cultural and historic roots of cosmology would have to include a look at the period of time known as the Renaissance. It is during this age that we find the beginning of science as we know it today. However, the foundation upon which the scientific method rests is very different from the way it was later practiced. This foundation grew out of a system of ancient knowledge that came from Egypt.

This may seem a little strange at first, especially since I haven't mentioned Egypt in my discussion of ancient cultural cosmology. But it will become very clear shortly. The greatest "scientific" minds of the Renaissance understood the world based on a body of knowledge that was unknown to most people. This chapter will explore that hidden knowledge.

While the study of philosophy might seem obscure to some people, *esoteric* philosophy would seem even more obscure. You can read about well-known philosophers, their ideas, and theories from ancient times to the present day. This is not what esoteric philosophy is, that would be considered *exoteric* philosophy. And this is true of many areas of study. Within religion for example, you can learn that all religions have their esoteric and exoteric sides. You can also apply that to history, science, psychology, economics, and mathematics to name a few. Let's see just what makes esoteric philosophy esoteric, and how this relates to the inquiring minds of the Renaissance.

On a fundamental level, philosophy or a philosopher seeks to understand the underlying truth or principles of the topic or thing they are investigating. Over the years philosophers have developed their own interpretation and understanding of answers to questions they've asked. Some are unique, some are reinterpretations of previous ideas, and some are a continuation of theories that came before them. For example, in Chapter 2, "The Music of the Spheres," you learned how Pythagoras took what he had learned in his travels and put together his theory of the music of the spheres.

Now esoteric philosophers—or another name by which they are known, *esotericists*—would take the process of looking at an underlying principle one step further. They would examine a number of philosophical theories about the structure of the universe and try to find a common thread to all of them.

Universal Constants

The word **esoteric** simply refers to something inner or within, but it has taken on more of a mysterious and arcane meaning over the last few centuries. It is often associated with secret societies, inner groups of initiates, and disciples who follow hidden doctrines and teachings. **Exoteric** is essentially just the opposite. It is what is found outside and usually pertains to things and ideas held by mainstream society or the general public.

Let's use an example from another area, such as religion. Instead of claiming that only one religion contains the truth or is the only true religion, an esotericist would study many religions and look for spiritual principles that are common to all and transcend the truth of any one religion.

Now, that's not all that esotericism is about, but this process does lie at the core of what an esotericist is trying to do. And what is that? They are seeking to know what lies hidden or below the surface, with the idea that there is a greater understanding to be achieved by seeing the unity in all things rather than what makes each one different from the other. To put it in their terms: the whole is greater than the sum of the parts. Understanding individual, particular things and then seeing what they share or have in common helps them to arrive at universal truths.

Universal Constants

An esotericist has sometimes also been called an occultist or a mystic. Some people use these words interchangeably. It really depends on the area of study each is seeking to understand; they also approach it from different ways. Generally speaking, the esotericist or occultist uses the mind to gain understanding, while the mystic tries to obtain the same knowledge through direct emotional experience.

Hermes and the Hermits

Besides being a method of inquiry in the search for universal truths, esoteric philosophy can also be a body of knowledge. The body of knowledge covered here is the esoteric philosophy that had the greatest impact on the Renaissance, which we'll discuss in Chapter 4, "Transitions in Thinking." This collection of knowledge is known as *Hermetic philosophy,* and it has its beginning in ancient Egypt.

Universal Constants

Hermetic philosophy is based on a collection of teachings and writings that originated with a legendary figure, Hermes Trismegistus, or the Thrice Great Hermes. This name is the Greek equivalent of the Egyptian god Thoth, the god of wisdom. Thrice Great, or having three parts, is usually interpreted as being the greatest philosopher, greatest priest, and greatest king. He is attributed with having written 36 books on theology and philosophy and 6 books on medicine.

The Egyptian god Thoth is said to have lived sometime in the very ancient past. He is described sometimes as being a god, half man and half god, or just a man. That's the problem with being a legendary figure, nobody knows for sure what you are. As a man he is said to have lived around 2700 B.C.E. As an Egyptian god, he played a role in creation myths, and he was the principle pleader for the human soul at the judgment of the dead. He is also considered to have introduced writing, mathematics, astrology, and alchemy. So in a nutshell, Thoth was seen as the source of sacred knowledge and of all sacred books in Egypt.

The Greek god Hermes became best known as the swift messenger of the gods. But he was also associated with invention and the agile movement of the mind that goes to and fro, like a messenger, connecting humans to the gods, exchanging ideas and commercial goods.

The Egyptian Hermeticists joined Thoth with the Greek Hermes to produce a new *archetype*, Hermes Trismegistus, who combined elements of both. Hermetic philosophy, or Hermeticism as it is often referred to, was one of the many products of the meeting of Greek and Egyptian cultures. It combined theology, philosophy, and various spiritual practices from both civilizations. Its origin probably occurred in or around Alexandria, Egypt, when it was the cultural capital of the Mediterranean under the Romans.

Universal Constants

An **archetype** is the term used to describe a symbol or mythic figure that is the original pattern or model from which all other similar things are made. In psychology it also represents inner psychological structures that are universal in nature.

The Emerald Tablet

The *Emerald Tablet* of Hermes Trismegistus is considered to be the cornerstone of Hermetic philosophy. It is the key text of Hermeticism. But there is much about it that is unknown. No one knows when this work was written, and as far as it being written on an emerald, well there is no known emerald large enough to hold the text. It is, however, considered to be the earliest of all Hermetic texts, possibly written by Thoth himself.

The original amount of Hermetic writings was thought to be considerable and a great portion of it was housed in the library of Alexandria. Many of these were lost during the systematic destruction of non-Christian literature that occurred between the fourth and sixth centuries C.E. However, the most fundamental writing, the *Corpus Hermeticum*, which contains the *Emerald Tablet*, still exists.

I thought you might enjoy seeing a translation of the *Emerald Tablet,* so here it is. But while it's not long, the language is very symbolic. Many alchemists believe it holds the secret of the philosopher's stone. If you've read *Harry Potter,* you know what I'm talking about. If you haven't, the philosopher's stone was the goal of all alchemists. It brought spiritual enlightenment and worldly wealth. Anything that it touched turned to gold.

Universal Constants

The **Corpus Hermeticum** dates from around the second century C.E. onward. It is a col-lection of 17 short Greek texts, some of which are works of Neoplatonic philosophy (new works based on Plato's ideas), Gnostic writings (such as those found in the Dead Sea Scrolls), and Cabalistic material (the mystical and esoteric doctrines of Judaism). Some of the *Corpus Hermeticum* was known to have been written by medieval thinkers, but the entire work was not translated into Latin until 1460 C.E.

I speak not of fictitious things, but of what is true and most certain.

What is Below corresponds to that which is Above, and that which is Above corresponds to that which is Below, to accomplish the miracles of the One Thing. And just as all things have come from this One Thing, through the med-itation of the One Mind, so do all created things originate from this One Thing, through Transformation.

Its father is the Sun; its mother the Moon. The Wind carries it in its belly; its nurse is the Earth. It is the origin of All, the consecration of the Universe; its in-herent Strength is perfected, if it is turned to Earth.

Separate the Earth from Fire, the Subtle from the Gross, gently and with great Ingenuity. It rises from the Earth to heaven and descends again to Earth, thereby combining within Itself the powers of both the Above and the Below.

Thus will you obtain the Glory of the Whole Universe. All Obscurity will be clear to you. This is the greatest force of all powers, because it overcomes every Subtle thing and penetrates every Solid thing.

In this way was the Universe created. From this comes many wondrous Applications, because this is the Pattern.

Therefore am I called Thrice Greatest Hermes, having all three parts of the wis-dom of the Whole Universe. Herein have I completely explained the Operation of the Sun.

Some of what the tablet is referring to is a basic idea found in Hermeticism, which is that correspondences exist between the world of the divine intellect and the created world. Knowledge is power for the Hermeticist—power over nature, himself, and his destiny.

The Perennial Philosophy

Hermeticism is part of what is called the Western esoteric tradition. This tradition embodies a set of spiritual teachings known as the perennial philosophy, the wisdom tradition, or the ageless wisdom. These esoteric teachings have been passed down through the centuries from generation to generation. Some scholars believe it is the inner driving force behind the flowering of the arts and sciences in many ages and forms much of the inspiration to be found in the truest aspects of the world's religions.

In Chapter 1, "Once Upon a Time," you were introduced to the concept of syncretism. This process occurred within Hermeticism from its inception to the present day. While the main principles of Hermeticism were firmly established during the first millennium C.E., it continued to draw upon other sources within the perennial philosophy and in turn also passed on its knowledge to other major Western esoteric movements. I'd like to provide you with a brief overview of some of these.

Black Holes

One of the greatest tragedies of all time was the destruction of the Alexandrian library. The collection of manuscripts was probably the largest in the Western world, numbering close to half a million. Around the third century C.E., the dawning Christian movement did not look kindly upon so-called pagan activities, and under Emperor Aurelian the greater part of the library was destroyed. In 400 C.E. Theophilus, Bishop of Alexandria, ordered many other books, which had been hidden away in the museum nearby, burnt. In 415 C.E. the brilliant Hypatia, the last head of the Library/Museum complex, was murdered by a group of monks. And in 642 C.E. the final destruction took place when Arabs conquered Egypt and used what was left of the scrolls to heat their baths. Just think of all of the incredible knowledge about the ancient world that was lost because of religious intolerance, ignorance, and fear.

Fruits of Knowledge

There are many branches on the tree of understanding. Some have died and withered, while others have blossomed and even produced fruit. Here is a list of some of the main branches that have survived pruning and are with us today:

- ➤ Rosicrucianism
- ➤ Freemasonry
- ➤ Cabalah
- ➤ Ancient mystery religions
- ➤ Gnosticism
- ➤ Esoteric Christianity
- ➤ Wicca and Neo-Paganism
- ➤ The Grail quest
- ➤ Theosophy
- ➤ Alchemy

These 10 paths are just some of the main branches of the tree of the Western esoteric tradition. There is a tremendous amount of information available on all of these in bookstores and on the Internet. Space doesn't allow for me to cover all ten of them, but I would like to provide you with an overview of the first three. The last one, alchemy, deserves an in-depth discussion because of its direct influence on Sir Issac Newton, whom we'll discuss in Part 2, "Science Comes of Age, but Isn't 21 Yet."

Rosicrucianism

Like many branches of a tree, the ten paths of the Western esoteric tradition have various smaller ones that have grown in different directions. While they still have the same source, there are differences in how they operate, the rituals that are practiced, and the way in which they interpret their teachings. For example, some Rosicrucian groups trace their history back to the Pharaoh Akhnaton (c. 1350 B.C.E.), considered to be the

Cosmonotes

Interest in the Western esoteric tradition has grown a lot over the twentieth century and especially during the last 25 years. The many different schools of thought have much to offer to those individuals seeking a deeper understanding to many of life's questions. Who knows what you might find out? The Shadow knows! (Eerie laughter in the background.)

Cosmonotes

If you ever get a chance to go to San Jose, California, visit the Rosicrucian Museum located there. It's designed to look like an Egyptian temple on the outside and houses a great Egyptian exhibit and an incredible library. However, the library is only open to members. But it's really worth seeing to learn about Rosicrucianism and ancient Egypt.

traditional master of the Ancient Egyptian Brotherhood. Others see their foundation beginning in the European Middle Ages, with it becoming very popular in the seventeenth-century Germany. This was due to the publication of a Rosicrucian Manifesto, the *Fama Fraternitatis* by a man known as Christian Rosenkrutz.

In either case, the Rosicrucian Order is known as the Ancient Mystical Order Rosae Crusis. The exoteric name "Rosicrucian" is derived from the Latin words Rosae or Rose, and Crux or Cross, giving you Rose Cross. But the word also has its symbolic, esoteric meaning and is explained within the context of membership.

Rosicrucianism has a detailed cosmology that draws on the teachings of many other esoteric traditions. It was this cosmology that attracted many European scientists, philosophers, artists, writers, and theologians of the seventeenth and eighteenth centuries. Some of the better-known ones influenced by these ideas were William Shakespeare, Sir Francis Bacon, Marie Corelli, Benjamin Franklin, John Dee, and Sir Isaac Newton.

Freemasonry

The history of Freemasonry can be broken into three periods: the legendary, the medieval, and the modern periods. The legendary period is called legendary because there are few Freemasons who consider it actual historical truth. As legend has it, the Masonic fraternity began with the construction of King Solomon's Temple. It was such a vast undertaking that a new form of organization was required to make sure it was completed correctly. This development placed stonemasons and architects into various grades and classes, similar to those found in present day Masonry.

In medieval Europe, the stonemasons were the ones who constructed the Gothic cathedrals. The term "free" in Freemason indicated that the Mason was not bound to the land as a serf, but was free to travel the countryside. So the word Freemason is sometimes used in place of Mason, although the latter is more common. When the building of the cathedrals ended and the Renaissance began, it is thought that the Masons wanted to maintain their organization by admitting new members to discuss the philosophical and more esoteric teachings of the guild. This is the accepted period of time for the origin of Freemasonry.

There is another theory for the medieval origin of the Masons. It has been suggested by some scholars that the Masons are descendants of the *Knights Templar*. Because of political and religious persecution it is thought that they preserved their fraternity by disguising it in the form of Freemasonry.

Universal Constants

The **Knights Templar** were a powerful and wealthy order of knights that had their origin during the Crusades. They were at one time the protectors of the Temple in Jerusalem, the Knights of the Temple. After becoming very wealthy over the years and owning large estates with castles, they were seen as a threat to the absolute power of the King of France and the Pope at the time. So during the fourteenth century many were tortured and put to death. On Friday, April 13th, 1307, Philip IV King of France put the knights on trial, and ever since then Friday the 13th has been considered unlucky.

The last part of Masonry that I want to touch on is the role it had in early American history. Many of the Founding Fathers were Masons, including some who signed the Declaration of Independence. Any guesses? Well, to name a few you have: George Washington, John Hancock, Thomas Jefferson, John Adams, and Benjamin Franklin. And have you ever looked closely at a one-dollar bill? It's loaded with Masonic symbols. Who would have guessed?

Cosmonotes

When you get a chance, check out the Gothic cathedrals in Europe. If you can't go in person, then look at some good pictures or watch a travelogue. What you'll see are Masonic symbols incorporated and sometimes innocently hidden in the facade of the building. Little did the Church know or understand the meaning of many of the symbols that the Masons were adding to the cathedral as it was being built.

The Cabalah

According to tradition, the Cabalah is the secret wisdom of God, first given to Adam and Moses and handed down over the centuries through oral tradition. The earliest written cabalistic text is the *Sepher Yetzirah,* or Book of Creation, which dates to sometime between the third and sixth century C.E. At the heart of all Cabalistic teachings is the "Tree of Life," shown in the following picture.

The Cabalistic Tree of Life.

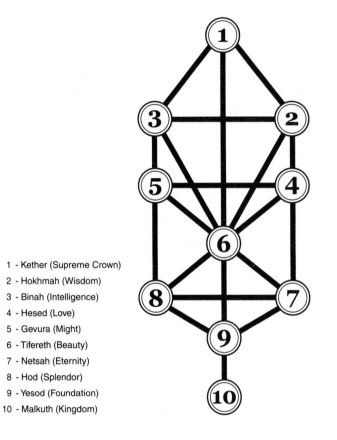

1 - Kether (Supreme Crown)
2 - Hokhmah (Wisdom)
3 - Binah (Intelligence)
4 - Hesed (Love)
5 - Gevura (Might)
6 - Tifereth (Beauty)
7 - Netsah (Eternity)
8 - Hod (Splendor)
9 - Yesod (Foundation)
10 - Malkuth (Kingdom)

Although the Cabalah originated in Judaism, it has been adapted to a number of other teachings within the Western esoteric tradition. There are Christian Cabalists, who have incorporated the teachings of the Christian mystic, Jacob Boehme. There is also the Cabalah of the occult, ceremonial magic groups such as The Order of the Golden Dawn. This group was at its height during the late nineteenth century with members such as Dion Fortune, Aleister Crowley, Israel Regardie, W. B. Yeats, and S. MacGregor Matthews.

The whole body of Cabalistic teachings is extremely large. It has correlations to cosmology, mythology, psychology, astrology, numerology, dream-working, theosophy, and the tarot. If you look at the picture of the Tree of Life and count the number of lines connecting the spheres, you will come up with a number equal to the number of *major arcana* in a tarot deck.

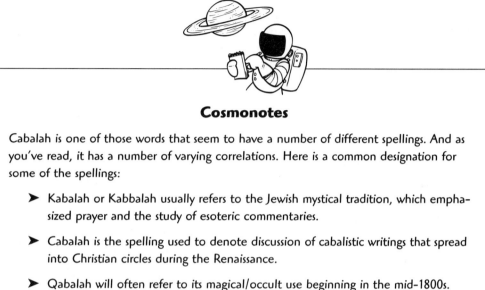

Cosmonotes

Cabalah is one of those words that seem to have a number of different spellings. And as you've read, it has a number of varying correlations. Here is a common designation for some of the spellings:

➤ Kabalah or Kabbalah usually refers to the Jewish mystical tradition, which emphasized prayer and the study of esoteric commentaries.

➤ Cabalah is the spelling used to denote discussion of cabalistic writings that spread into Christian circles during the Renaissance.

➤ Qabalah will often refer to its magical/occult use beginning in the mid-1800s. During this period many scattered elements were brought together and synthesized into a system of correspondences.

Why All of This Is So Important

You've now spent the last 10 pages reading about esoteric philosophy. While it's been interesting, you may be asking yourself what it all has to do with theories of the universe. Hopefully along the way you've developed a sense of how all of these systems of thought contained their own cosmology. Their importance as systems of thought will be one of the key factors influencing the minds of the Renaissance. The Hermetic tradition and its emphasis on the goal of perfection helped to raise humanity out of the Dark Ages into the new.

The Renaissance, or rebirth, was a period of time after the Middle Ages when people began to change their perception of how they understood the universe and especially their relationship to the world in which they lived. It was a period of tremendous flowering in art, philosophy, and science. But this great rebirth and its forward

Universal Constants

The **major arcana** of a tarot deck comprise the 22, what could be called trump cards. The rest of the deck is made up of four other suits, which have a correlation to playing cards. The four suits are: pentacles or diamonds, cups or hearts; swords or spades, and wands or clubs. Far from being just a popular form of fortune telling, it is a symbolic system that you'll learn more about in Chapter 24, "Symbolic Systems and the Self-Aware Universe."

movement derived its drive and emotional impulse from looking backward, into the past, not from any ideas of looking forward.

This was a time of rediscovery, of the resurrection of classical learning. And it was also a return to the ancient teachings of Hermeticism. In 1460, one of the most influential people of the Renaissance, Cosimo de Medici, got his hands on an almost complete manuscript of the *Corpus Hermeticum*. He handed it over to Marsilio Ficino, a brilliant young philosopher who was running Cosimo's Platonic academy. Marsilio stopped his work on Plato's dialogues and was told to translate the new manuscript. After that the rest is history.

The discovery of the Hermetic texts created a sensation in the Florentine intellectual circles. From there it spread like wild fire throughout Europe to the scholars, historians, artists, philosophers, theologians, and soon-to-be scientists. It's difficult to put into words the degree to which these teachings would impact the Western intellectual tradition for the next three hundred years. The most inspiring belief that was to have the strongest influence was that humanity could remake its own destiny.

The Least You Need to Know

➤ Esoteric philosophy is the study of the deep universal principles that govern the cosmos, society, and the individual.

➤ Hermetic philosophy is based on the teachings of Hermes Trismegistus, or the Thrice Great Hermes.

➤ The Western esoteric tradition incorporates writings, teachings, and practices from many different spiritual traditions and is also called the perennial philosophy or the ageless wisdom.

➤ The influence of Hermetic philosophy was very widespread during the Renaissance, affecting people's beliefs about the universe and themselves.

Chapter 4

Transitions in Thinking

> ## In This Chapter
>
> ➤ Life in the Middle Ages
>
> ➤ The transition from medieval to Renaissance thinking
>
> ➤ The birth of humanism
>
> ➤ Hermetic philosophy in the Renaissance

The stage is now set for you to see some important changes that were taking place between the fourteenth and seventeenth centuries. As you saw in the last chapter, the classical knowledge of antiquity, especially that of Hermeticism, was making its way into the consciousness of the Renaissance mind. At this point I think it's important to explain why.

The transition from the Middle Ages to the Renaissance was one of the most significant occurrences in the history of Western civilization. It was a change that took place on a number of levels within the entire European culture. From philosophy and art to science, politics, economics, and theology, no aspect of human endeavor was ever the same after the Renaissance. At the core of this transformation was the way in which people understood their place in relationship to the world around them. So let's take a look at how this understanding changed.

Simple Minds, Simple Pleasures

Probably one of the most unusual characteristics that defined the medieval mind was something that was invisible and silent—a total lack of ego, or self-identity. That's a hard concept for us in the twentieth century to grasp. We have known no other way of being in the world other than through our sense of self. And history is always viewed through the eyes of the present looking into the past. The real difficulty in truly understanding the past is not to interpret it with the mind of today.

Historical novels and the movies of Hollywood project our perceptions onto the past and provide a reality in which a story can be told, but at the same time the lack of a true portrayal of the past makes it appear as though very little has changed in the way people saw themselves. So let me draw this veil of illusion aside and try to give a brief, but maybe a little more accurate, picture of life in the Middle Ages.

As I mentioned, the lack of a sense of self was commonplace during this time. Let me give you an example to illustrate why this was so. The great Gothic cathedrals, which are probably the most treasured legacies from this period, usually took three or four centuries to complete. Canterbury Cathedral in England took 23 generations of workers to build; Chartres Cathedral in France, which was a former *Druidic* center, took eighteen generations. Yet who were these people? We know nothing about them. Their mission was to glorify God. For them, who they were really didn't matter.

Although most noble families had surnames, that accounted for less than one percent of the entire population. Everybody else, which was roughly about 65 million people, were known as Franz, Jacque, John, or John's son (Johnson), John's daughter, or John's wife. People very seldom left the village of their birthplace, so stronger means of identification wasn't necessary. Nicknames also worked well. One-Arm, Red-Nose, or Blondie would have easily identified who you were.

Mindwarps

A major portion of the population of Europe was illiterate during medieval times (750–1350 C.E.). There were no newspapers or magazines to inform people of great events, and even if there were nobody would have been able to read them. An occasional pamphlet might reach you, but these were usually theological, and like the Bible, were written in Latin, which nobody understood outside of the clergy.

Universal Constants

The word **Druidic** refers to the priestly caste of the ancient Celtic people of the British Isles, the Druids. Stonehenge is one of their most famous sacred sites. However, they weren't confined to ancient Britain. Archeological evidence shows that they were widespread throughout Gaul, the pre-Roman lands of what is today France and Belgium. It was not uncommon for the bishops of medieval Christianity to have churches or cathedrals built on ancient sacred sites of the pre-Christian religions of Europe. At times it made conversion to Christianity somewhat easier if both religions shared the same sacred spot.

As it was for individuals, so to was it the same for the villages in which they lived. If you were to have gone off to war or gotten lost traveling in the countryside, it would have been impossible to ask anyone where your village was, for it was indistinguishable from any other. A landmark such as a hill, an unusual tree, or maybe a creek could only have identified it, if there had been one. Each hamlet was inbred, isolated, and totally unaware of the world outside of its familiar boundaries.

The one main focal point for all of these people was the Church. Through it they were baptized, attended Mass, married, and received last rites. The idea of being informed of great events would not have occurred to them, nor would they have even thought of voicing any response to such events. What you did, how you did it, what you thought and believed were all strongly governed by religion and totally accepted by the people.

There was something else missing in the Middle Ages that we live our lives by today—awareness of time! It's very easy for us to know the current month, day, and year. We are almost instinctively aware of the past, present, and future. It's unthinkable for us not to even know what century we're living in, but this was indeed true for medieval

Mindwarps

Identities would eventually become necessary and oftentimes people would adopt the surname of the local lord. This practice occurred in America as well. After the Civil War and emancipation, many ex-plantation slaves took on the name of their previous owner. Another common practice was to take the name of an honest occupation such as Taylor, Smith, or Miller.

41

people. There was absolutely no need for them to know what century it was. There were no watches, clocks, or calendars. Life for them revolved around the cycles of nature, religious holidays, and harvest time. Each family's generation followed another in an endless, timeless blur. While the outside world went through its popes, kings, and wars, and the little communities suffered hardship and natural disasters, they always managed to recover.

Any idea of introducing something new was riddled with suspicion. Invention or innovation was inconceivable, especially with the strong possibility of being accused of anti-Christian activities and having to withstand ordeals to prove your innocence.

Life in the Fast Lane

If time travel were possible, I think we'd all avoid going back to medieval times. Doesn't sound like there was much going on. But things did begin changing as larger and larger communities began forming into cities and towns. This took place, as it did in ancient Mesopotamia, along trade routes. The traveling merchant of the Middle Ages would help establish permanent areas of business that attracted people from the small villages. Trades began to flourish and life started to become interesting and not so innocuous.

Before you knew it, skilled craftsmen were forming guilds, architecture began taking on other forms besides churches and cathedrals, and travel outside of the normal familiar boundaries became exciting. Change was definitely taking place, but still rather slowly. If it hadn't been for the Crusades, the Renaissance probably would never have happened. Of course there were other factors involved that also gave rise to the Renaissance, but it was the Crusades that put people in touch with a much larger world and reintroduced the knowledge of *classical civilization* into Europe.

Universal Constants

I've made reference to **classical civilization** a few times already, so it's probably a good idea to define exactly what that pertains to. The empire established by Alexander the Great spread Greek culture throughout the Mediterranean world and as far east as India. The Greek ideals present in Greek art, philosophy, architecture, and science influenced and altered every culture they came in contact with. This period of time is called the Hellenistic Age and is defined as one of the classical civilizations. When Rome conquered and established its empire, the second great classical civilization was born. The cultural achievements of these two empires are what are referred to as classical civilization.

In 1000 C.E. Europe consisted of stone fortresses on hills and muddy huts in cramped villages. Two centuries later, the world was transformed. This coincides with the period of the first four Crusades. Soldiers returning from the Crusades brought tales of the East and of marvelous fabrics and goods, especially spice, that caught the fancy of nobles and commoners alike.

Along with all of these fantastic things came, of course, knowledge. With trade routes being established with the Eastern Empire and the Islamic countries, Greek philosophy, Roman law, and the esoteric works of Arabic alchemists began making their way into the intellectual circles of the trade cities of Europe.

Cosmonotes

Nowhere in the medieval world was the classical tradition better preserved than in Constantinople. Today it is known as Istanbul. Although it was heir to the Roman Empire, the Eastern Empire was directly descended from classical Greece. There was an innate love for antiquity, because it represented the best of what humanity had achieved. It was this love for the past that would also become the obsession of Renaissance Europe. This is a very fascinating period of time and I would recommend reading more about it so you can get a deeper appreciation for the richness of the Eastern cultures at this time.

The Birth of the Renaissance Ideal

Medieval thinkers can be divided into two basic types. Those that believed that since God has spoken to us it is no longer necessary for us to think; and the other believing that divine law required people to seek God by the rational methods of philosophy. Both kinds proceeded from fixed premises. Fortunately for us, the first kind didn't make it very far.

What exactly changed when we talk about transitions in thinking from the medieval mind to that of the person of the Renaissance? For the thinkers of the Middle Ages, and there weren't many of those, they largely accepted that God had created the world. The main questions they would have asked are, what is the nature of God? and, what is the relationship between God and man? In other words, their curiosity centered around God, not the world.

Universal Constants

Humanism or humanistic philosophy had its specific origin with the writings of Petrarch around 1341 C.E. in Italy. He is one of the key figures in the transition from medieval to Renaissance thought. He had a passion for classical literature and sought to discover humankind's own earthly fulfillment. He has been given the title of Father of Humanism.

Universal Constants

The **Renaissance ideal,** very simply, was that knowledge about the world ultimately depended on our perception of it and not anything else. It didn't deny the existence of God, but just believing in his existence was not enough. The world that had been created by him needed to be understood by either the use of rational thought, or acute observation.

What really shifted, then, was the focus of curiosity. For the Renaissance person, investigating the world became the main point of interest. And as this process continued to spread, it became an insatiable thirst. Out of this need to observe and understand the natural world, the scientific method eventually developed. But that method was about two hundred years away. A few other transitions in thinking had to take place before science became the methodology of choice for examining the world.

The Good Life

With this change of focus from God to the observable world came a philosophy that reflected a new view of humankind as well. The medieval view of life as a vale of tears, with no purpose other than to prepare for salvation and the afterlife, gave way to what was viewed as a more liberating ideal of people playing important roles in this world. The new view was called *humanism.*

Humanism was not a denial of God or faith. It was a philosophy that separated from religious doctrine humanity's ability to fulfill itself. It developed an increasing distaste for dogma, embraced a more figurative rather than literal interpretation of the scriptures, and developed an attitude of tolerance toward all viewpoints.

Perfection Is the Key

When a new worldview captures our imagination, a rich outpouring of creativity occurs in all areas of life. The Renaissance was an age dominated by genius. The number of outstanding personalities was incredible compared to previous ages. With the freedom to pursue new areas of thinking and expression, many sought to embody what is called the *Renaissance ideal.* Individuals such as Leonardo Da Vinci, Michelangelo, Erasmus, and Luther defined this ideal for others to follow.

In conjunction with the rise of humanistic philosophy and the Renaissance ideal, which both focused on the dignity and intrinsic value of the individual, was another important idea: Human beings were both good and ultimately perfectible. And this idea of the perfectibility of human beings became the goal of Hermeticism, and generally, the goal of the entire body of esoteric philosophy. How does one achieve this goal? You can achieve it through any of the paths that the Western esoteric tradition has to offer.

The Hermetic Revival

You now have a pretty good overview of the shift that took place in thinking from the minds of the Middle Ages to those of the Renaissance. And as previously mentioned, along with all of the classical knowledge that was being embraced by the people of the Renaissance, the Hermetic writings of Hermes Trismegistus were also being widely read. Now, the reaction of the Christian establishment to these writings was surprisingly rather ambivalent. They were never condemned and were revered by many prominent theologians.

In 1593, Cardinal Patrizzi had an authoritative volume of the Hermetic books printed. He also recommended that these works replace Aristotle as the basis for Christian philosophy and be diligently studied in schools and monasteries. What an incredible turn Western culture might have taken had the Hermetic teachings replaced the Aristotelian theology of Thomas Aquinas as the central doctrine of the Catholic Church.

While many people embraced Renaissance Hermeticism, there were three individuals at this time who were responsible for its widespread influence. All three wanted to tie it into Christian doctrine. The first two, whom I will discuss shortly, did this very well; the third went a little too far and was burned at the stake.

Mindwarps

In the development of the main doctrines of the Catholic Church there are two individuals who played a crucial role. The first, St. Augustine (334–430 c.e.), molded Platonic philosophy into Christian faith. He did this so well that it would be impossible to remove Platonism from Christianity without tearing it to pieces. The second, St. Thomas Aquinas (1225–1274 c.e.), incorporated the teachings of Aristotle into his Christian philosophy. His work is the official Catholic philosophy and to this day is taught in Catholic schools and universities.

Marsilio Ficino

You met Marsilio at the end of Chapter 3, "A Brief History of Esoteric Philosophy." He was the individual who translated the *Corpus Hermeticum* into Latin and essentially made it available for everyone else to read. The Platonic Academy that he had set up with the help of Cosimo de Medici was the first school in Europe to begin incorporating Hermetic philosophy into the curriculum. Here is a summary of the major teachings that expressed some of the main Hermetic teachings:

➤ **As above, so below**   The basic idea of Hermeticism is that there are real correspondences between heaven, the divine world of the intellect, and earth. In Christian doctrine its equivalent would be the Kingdom of Heaven on earth.

➤ **A historical view of religion**   The wisdom contained in the Hermetic and Christian writings, rather than opposing each other, should be harmonized.

➤ **The practice of contemplation**   To encourage and experience the Divine Mind, with the goal of discovering one's own spiritual identity.

➤ **Platonic love**   A divine form of friendship between two individuals involved in contemplation. It seeks to mimic the love they both have for God.

Because of Ficino, Florence remained the center of humanistic and intellectual movements until his death in 1499. One of Marsilio's students took these basic premises even further to show the spiritual potential that lies within each of us.

Mindwarps

With the concepts put forth by Marsilio Ficino and Pico della Mirandola you can see the stirrings of early scientific methodology. With ideas like the unity of nature, the soul of the universe (remember Plato's world soul?), and the divinity of humanity, the pursuit had begun to search nature and unlock its secrets to prove that these ideas were true. Since earth reflected the divine laws of heaven, what better place to go looking than in the natural world?

Pico della Mirandola

The influence of Marsilio's academy and the Hermetic philosophy on Pico della Mirandola is evident in the opening paragraph of his most famous work, *Oration on the Dignity of Man,* which he delivered in Rome in 1486, at the ripe old age of 24. With his statement taken directly from Hermes, "What a great miracle is Man," he endeavors to show that at our true essence we are really gods who have become beggars. Armed with the works of Plato, the Gospels, the writings of the Church Fathers and most importantly, the texts of Hermes Trismegisus, he works to convince others of their innate goodness and divinity.

This supreme self-confidence is considered to be one of the legacies left to us from the Renaissance. After Pico, European peoples had the confidence to act upon the world and to control their destiny through knowledge. Unfortunately for Pico, he never attained what he sought, dying of fever at the age of 31.

Giordano Bruno

Although Marsilio and Pico didn't have any problems with the Catholic Church, Bruno ended up being executed. However, he was later considered a martyr for two causes; a defender of free thinking who questioned the dogma of the church, and as a supporter of Hermetic philosophy.

You see, from the fifteenth century until almost the end of the seventeenth century the Hermetic texts remained very popular. It was not until the *Protestant Reformation* and the ensuing religious strife that followed that these teachings began to lose their status. And unfortunately, Giordano came onto the scene at a bad time. The tolerance of the Catholic Church for anything other than Church doctrine disappeared, and you could easily find yourself burning at the stake.

Universal Constants

The **Protestant Reformation** occurred toward the end of the fifteenth century and the early part of the sixteenth. Although Martin Luther is often considered the instigator of it, rumblings were already occurring a hundred years before his break with the Catholic Church. Throughout Roman Catholic Europe, the huge body of clergy had amassed considerable wealth.

This, in turn, irritated secular governments, especially since the Church taxed the secular sector heavily. There was widespread resentment of central ecclesiastic authority and many people were getting fed up with Church dogma.

Bruno is most commonly remembered for his defense of the Copernican solar system. (If you're not sure what I'm talking about, don't worry, that's coming up in Chapter 5, "You Mean We're Not at the Center of the Universe?") He is seldom remembered for his support of Hermetic doctrines. In 1600 he was burned at the stake mainly for wanting to revive the ancient Egyptian, Hermetic religion. But his contribution to the Hermetic Renaissance is a central point and two aspects of his work should be pointed out.

Bruno's first contribution deals with his mastery of something called the art of memory. This was a method known to the ancient magi, revived by Renaissance scholars, and used in magical practices of modern esotericists. Bruno called it "magical memory," and as a central theme in his work it was an important practice. Essentially it was a powerful technique used to remember anything imprinted within the mind. Bruno adapted it to his magical projects, taking the divine images from the Hermetic books. He fixed these in his imagination, thus furnishing his inner world with a blueprint of the universe. By reflecting the universe in his mind, this Renaissance magus fulfilled Pico's injunction that humanity must embody the highest and greatest good. The doors of cosmology would swing open, and like Pythagoras, he could hear the music of the spheres.

The second aspect of his work, and the one that did him in, was his desire to break the hold of the Church over people's minds. He wanted to accomplish this by reviving the ancient Hermetic religion and, in doing so, replace the image of man put forth by the Church. For Bruno, the Christian Church was one of the agents responsible for the darkening of humankind's spiritual light, and the fall into forgetfulness of its divine heritage.

Bruno proclaimed that the new Copernican theory of a sun-centered solar system was evidence that the Hermetic revival had come. His use of Copernicus, coupled with his bombastic, aggressive personality led to his death. After years of bitter struggle against the papal forces, this would-be founder of the new hermetic age was burned alive by the Inquisition in 1600.

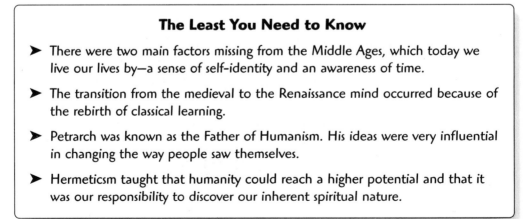

The Least You Need to Know

➤ There were two main factors missing from the Middle Ages, which today we live our lives by—a sense of self-identity and an awareness of time.

➤ The transition from the medieval to the Renaissance mind occurred because of the rebirth of classical learning.

➤ Petrarch was known as the Father of Humanism. His ideas were very influential in changing the way people saw themselves.

➤ Hermeticism taught that humanity could reach a higher potential and that it was our responsibility to discover our inherent spiritual nature.

Part 2

Science Comes of Age, but Isn't 21 Yet

The sixteenth and seventeenth centuries were very tumultuous times. The Renaissance had opened people's minds to a greater world. Religion no longer governed every aspect of society as it had in the Middles Ages. Freedom to think outside of strict religious paradigms brought long-held cherished beliefs about the inherent structure of the universe into question. The desire to know and understand the natural world without biblical reference was one of the factors ushering in the scientific age.

The accepted cosmology of the universe was turned on its head. As much as the Renaissance had impacted people's lives, very few were prepared for the revelation that displaced them from a prominent place in the universe. And as Humpty Dumpty couldn't put all the pieces back together again, so, too, was the comforting thought that God had created everything just for us and that we were the most important thing in the universe.

Chapter 5

You Mean We're Not at the Center of the Universe?

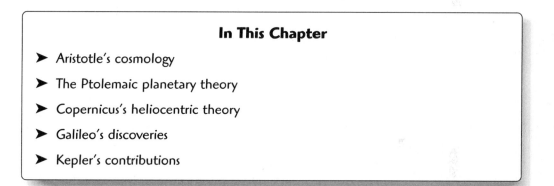

In This Chapter

➤ Aristotle's cosmology

➤ The Ptolemaic planetary theory

➤ Copernicus's heliocentric theory

➤ Galileo's discoveries

➤ Kepler's contributions

Imagine waking up one morning and finding out that everything you thought you knew about the world around you was wrong. Well, it didn't happen quite that quickly, it was more like 200 years, but it still threw everyone for a loop. New ideas, especially those that affect how we think of ourselves, are never easily accepted. As you will soon see, not only were these new ideas not accepted, but also the proponents of them put their lives in jeopardy by advocating them.

The first four chapters have given you some idea of the cultural, religious, and philosophical views of the universe. We've looked at its origin and structure from the viewpoint of some of the earliest civilizations. You may have also heard the music of the spheres and dabbled a little in esoteric philosophy. Now it's time to examine how these perspectives were about to change and why.

What You See Is What You Get

In Chapter 2, "The Music of the Spheres," you read about Pythagoras's music of the spheres and Plato's world soul and first cause. These philosophers described the structure of the universe through mathematical relationships. They had many ideas in common, but there are two that we need to focus on:

➤ The universe was a dynamic, living, organic structure that was perfectly ordered and based on musically mathematical proportions.

➤ The Earth was at the center of the universe and all of the *planets* revolved around it in perfect circles.

It was these two fundamental concepts about the universe that changed during the Renaissance. But there's a lot that happened before this shift took place. Let's begin by looking at another Greek philosopher that I've already mentioned—Aristotle.

Plato's most distinguished pupil was Aristotle (c. 384–322 B.C.E.), on whom Plato had a great influence. Aristotle was eventually hired to be a teacher of Alexander the Great, but if Aristotle, in turn, had a tremendous influence on Alexander, there is little evidence of it. It's a good bet that Alexander, who conquered much of the known world, had other things on his mind besides philosophy.

Aristotle was a very careful observer of nature and a brilliant theorizer. His ideas, like Plato's, influenced philosophy for centuries. Some 1,500 years after his death, he was considered the definitive authority on all subjects outside of religion, although he indirectly influenced Church doctrine through the philosophy of St. Thomas Aquinas. Remember him from the last chapter?

The fact that Aristotle was considered such an authority most likely impeded rather than helped scientific progress. In science, like in many other areas, you can't assume something is true just because some authority says that it's so, even if that authority is Aristotle.

So Aristotle, like Pythagoras, Plato, and just about everybody else, believed in a *geocentric* theory of the

Universal Constants

Planet comes from a Greek word meaning "to lead astray or wander." In Aristotle's system, the stars were fixed in the heavens, but there were other objects that seemed to move around or wander with no set destination. These were called planets. Do you remember what the seven sacred planets were?

Universal Constants

Geocentric is made up of two Greek terms: "geo," meaning earth, as in geology or the study of the Earth; and "centric" or center. Thus, you have earth-centered. So the geocentric theory of the universe just meant a theory in which the Earth was considered the center, and everything revolved around it.

universe. But Aristotle didn't believe it just because everyone else did; he believed it because that's what he observed—a crucial difference. And this is a key factor in what made Aristotle's ideas so powerful and widely accepted, the fact that they were based on actual observations.

Just think about it. If you didn't know that the Sun was stationary and that the Earth revolved around the Sun, what would your senses tell you? From the time the Sun rises (which implies that it's moving), until it sets, you see the Sun moving across the sky. There is no sensation of the Earth turning on its axis or moving in orbit around the Sun. So why question something that seems so apparent? And that's just it, no one did.

In Aristotle's explanation of how the universe was structured, there was already a change in its innate quality. There was no sense of it being a living, organic process based on mathematical proportions equivalent to musical harmonies. Aristotle combined the use of rational thought with observation to explain the structure of the universe. And these are two of the three methods used in science. The third, the idea to test the observation, wouldn't come until the Renaissance, but because he used these two methods, Aristotle is sometimes considered the first scientist.

Back in Chapter 1, "Once Upon a Time," I said that there were two ways in which people have tried to understand the origin and structure of the universe—numerical systems and observation of the stars. We will come back to these two methods throughout this book, because regardless of the time period or the culture, these two methods are the main themes underlying all the systems of thought pertaining to cosmology. Aristotle was no different. Even though his views differed from his predecessors, he still used observation and numbers to explain his version of how the universe was structured.

In Aristotle's version, the universe was divided into two domains, the heavenly and what he called the sublunar or earthly realms. The heavenly domain was a complicated system of 56 interconnecting spherical shells, each of which derived its movement from the other shells linked to it. The outermost spherical shell was the prime mover, which started all other shells moving. It is reminiscent of Plato's first cause, although his demiurge is nowhere to be found in Aristotle's version.

The earthly realm was composed of four substances: earth, air, fire, and water. These four sublunar elementary materials interacted with each other to form all of the material objects found in the natural world.

Walk Like an Egyptian

Let's leave Aristotle and move ahead about 400 years to around 150 C.E. We now meet Claudius Ptolemy (c. 100–170 C.E.), an Egyptian astronomer. He wanted a theory of cosmology that not only accurately predicted the position of the planets, but also provided a true image of the cosmos.

From the close of Greek antiquity to the astronomical changes during the sixteenth century, Ptolemy's astronomy and cosmology, very nicely combined with Aristotle's physics, were almost universally accepted in the West. This was true wherever Greek astronomy was preserved, from Islam to medieval Europe. The medieval astronomers and cosmologists who wrote in Greek, Arabic, Latin, or Hebrew were all heavily influenced by Ptolemy's cosmology. Each culture modified his system in minor ways, often to make them more consistent with religious and philosophical requirements. But in any case, for the great majority of time, his was the reigning cosmology.

Cosmonotes

If you want to read about some of the first and original ideas about the structure and composition of the universe, read the writings of the pre-Socratics—you know, the philosophers who came before Socrates. Among them you'll find Empedecles, the philosopher who said that all of the objects in the natural world are made up of the four substances: earth, air, fire, and water. The alchemists of the Renaissance used this basic composition, not literally, but symbolically.

There is very little known about Ptolemy's life except that his astronomical observations took place between 127 and 141 C.E. in Alexandria, Egypt. But he did write an important work called the *Almagest,* which is a treatise on his mathematical astronomy. The main subject of his theory is the motion of the sun, moon, and planets. He introduces geometrical models to explain these motions and develops what's called a *planetary theory.*

What was really important about Ptolemy's planetary theory was that it brought together the two methods we keep coming back to, numerical systems and observation, in a whole new way. His mathematical explanation for the structure of the universe was in excellent agreement with observed celestial motion. Pythagoras's and Plato's systems, on the other hand, explained the structure of the universe using numerical systems and observation, but their systems didn't use the two methods to explain what was actually observed in the sky. In other words, their systems didn't account for the movement of the Sun, the Moon, or the planets as they traveled across the heavens.

Universal Constants

A **planetary theory** is a mathematical scheme that can be used to calculate the position of a planet, the Sun, or the Moon in the sky for any desired date.

And while Aristotle introduced the important roles of observation and reason, his mathematical system wasn't very good at accounting for the motion of the planets either. It was mostly just an awkward theory. As a matter of fact, the Ptolemaic planetary theory was held in such high regard by astronomers in Rome, Egypt, Arabia, and Greece that few people paid much attention to the theories of his predecessors. So, it is with Ptolemy's theory that we can see a change in people's perspective and the beginning of what would become the *scientific method.* Why? Because observation was being supported by a mathematical system and the two methods helped to verify one another.

Universal Constants

The **scientific method** is a process in which a theory or hypothesis is developed based on observation or the collection of data. Then the theory is tested through experimentation or the application of mathematics. The end result should verify the original theory. If it doesn't, then it's time to come up with a new theory.

Besides the *Almagest,* Ptolemy also wrote about geography, optics, philosophy, the art of sundials, and a work on music theory. In keeping with Pythagorean tradition, he gave various features of the motions of the planets' musical meaning. For example, longitude or moving from east to west corresponded to pitch. He also wrote works on astrology and its correlation to planetary motion. So while there is an addition of a mathematical system to verify observation, musical theory and ancient astrological ideas still played an important role in describing a complete cosmology.

One last idea of Ptolemy's that is worth looking at before we move on is his idea of no empty space in the universe. Similar to Aristotle's spherical shells, but much simpler, Ptolemy worked out a system of eight nested spheres, each one accounting for the motion of one of the seven sacred planets, with the last shell corresponding to the fixed stars. The entire universe is made up of these eight shells, with no room or space in between.

The Copernican Revolution

For the next 1,000 years, the Ptolemaic system was the cosmology accepted by almost everyone, or at least those people who cared. Remember that people in the Middle Ages didn't hang out at their local bars discussing the pros and cons of various cosmological theories.

In 1543, Nicolaus Copernicus's (1473–1543) *On the Revolution of Heavenly Spheres* was published. This book set forth his idea of a *heliocentric* planetary system. And while this was a truly new and revolutionary theory, it only composed about 5 percent of the text. The rest of his work is essentially just a rewrite of Ptolemy's *Almagest.* Copernicus departs from some of Ptolemy's technical details, but for the most part follows him. He relies on his observations, emulates his methods, and often adopts

Universal Constants

Heliocentric means that the Sun is at the center. *Helio* is the Greek word for "sun." So heliocentric planetary system just means a sun-centered theory of the universe, as geocentric meant earth-centered.

his mathematical parameters. So even until the end of the sixteenth century, astronomers, while disagreeing with some of Ptolemy's ideas, still based their explanations on much of his material.

In case you didn't notice, the year that Copernicus's book was published was also the year of his death. It was thought that he wanted to wait until after his death to have the book published, because he feared possible repercussions from the Catholic hierarchy. There are variations to this story. Some historians place him publicly supporting his theory in 1540, three years before his death. Others propose that the book was almost never published, and it was only through the insistence of a friend and the pressure from a bishop that Copernicus finally did. In either case, he dedicated the book to the pope at the time, Pope Paul III.

Mindwarps

Even though Copernicus is attributed with having developed heliocentric planetary theory, he wasn't the first one to really think of it. In the first half of the third century B.C.E., Aristarchus of Samos postulated a heliocentric theory. That's 1,800 years before Copernicus. There were also a few Arabic astronomers who questioned the geocentric system they inherited from the Greeks and Ptolemy. Copernicus just happens to be the guy who gets the credit.

The Problem with Easter

Prior to the publication of Copernican heliocentric theory, the long-held system of Ptolemy was already beginning to be questioned. Remember that when it was originally put forth, one of the hallmarks of the theory was that it was the first one to bring together observation with mathematics. However accurate this description was initially, over time something appeared to be seriously wrong. Centuries of celestial observation began to see weaknesses in the Ptolemaic system that couldn't be accounted for.

Cosmonotes

An interesting note to the geocentric and heliocentric theories is that although we have adopted the heliocentric as representative of the actual structure of our solar system, it makes no difference which one is used for accurate astronomical prediction. In other words, whether the Sun goes around the Earth, or the Earth goes around the Sun merely reflects the choice of a frame of reference. Sun-centered theories are, therefore, not intrinsically any more accurate than are earth-centered theories. The accuracy of the theory depends on what you do with it.

Calendars are an indispensable part of every society. They are like huge timepieces that click away the days, weeks, and months contained within a defined period of time. That defined period is based either on a solar cycle or a lunar cycle. In both cases they are based upon celestial movements. Our calendars are a combination of astrological and astronomical systems resulting from observations made about the relative position of the Sun, Moon, and planets to us here on earth. So there must be a direct correlation between what is observed in the skies and what date, or time of year it is here on earth.

We know today that it takes the Earth 365 days, 5 hours, 48 minutes, and 46 seconds to make one annual orbit around the Sun. The ancient Egyptians, whose calendar was adopted by Julius Caesar (and modestly called the Julian Calendar), calculated the year to be 365 days and 6 hours. Oops, that's 11 minutes too long. That doesn't seem to be too big of a deal, until over time, the 11 minutes are added to each year. Before you know it, 1,500 years later, the calendar is now off by 10 or 11 days. Ptolemy's mathematical system of prediction was no longer matching the dates on the calendar. But if that wasn't enough, Easter was hopping farther and farther away from the day of the *vernal equinox,* its fixed date of March 21.

The Church had set Easter at March 21—the vernal equinox, the first day of spring. By the fifteenth century, the equinox was occurring on the calendar around March 11. Something was seriously wrong, which meant Ptolemy's complex system wasn't working.

Displaced Humanity

How do you think everyone reacted to this new theory? If aliens were to land on earth tomorrow, how would you react? Well, this would be about the equivalent.

Universal Constants

The **vernal equinox** is also called the spring equinox. Vernal comes from the Latin word *vernalis*, which means "spring," and equinox means "equal night." In other words, at both the spring and autumn equinox, the length of day and night is equal.

Initially the Church didn't denounce Copernicus's publication, but by 1616 it was officially condemned as heretical.

How could Aristotle, Ptolemy, and the Bible be wrong? Unthinkable! God had placed humanity on a world at the center of the universe. We were the most important things created. Churchmen were convinced of our unique placement. The world was a static and unchanging place since it was created. How could we become displaced onto just another planet orbiting the sun? We would lose our self-importance and that wasn't going to happen.

There were a couple of common sense arguments to oppose this theory as well:

➤ Gravity was unknown at this time, so the idea of things falling to the ground was based on the Aristotelian notion that the Earth was at the center of the universe. If the Sun were at the center, objects would fall toward the Sun, not the Earth. Duh!

➤ If the Earth were moving, we would feel the wind constantly blowing on us. After all, whenever you move, like in a carriage, or even just running, don't you feel a breeze?

In the end, there was no proof that Copernicus's new theory was right. It was just a theory. But proof would be coming in less than 100 years. Galileo was about to use his telescope to show that the heliocentric theory was correct.

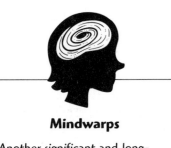

Mindwarps

Another significant and long-held belief changed around the time of the acceptance of the Copernican system. Most people believed that the Earth was a flat disk made up of Europe and Asia. But with the advent of ocean exploration, the realization came that it was a sphere covered mostly with water.

Galileo Scopes It Out

Most of Galileo's evidence in support of the Copernican theory was published in a book called *The Starry Messenger*. It was this publication that first got him accused of religious heresy. So in the hopes of getting other people to see what he saw, he invited a group of

Jesuit priests to gaze through his telescope. All declined except for one, who still couldn't believe what he saw. He was convinced that the telescope altered reality and that without it everything was still the same.

Mindwarps

Galileo Galilei (1564–1642) is most often associated with the invention of the telescope. He actually didn't invent it; that was done by a Dutch spectacle maker by the name of Hans Lippershey in 1600. What he did do was to improve on its design by increasing its power by more than three times over anything built before. His 30-power telescope was the strongest anyone had ever produced. With this new instrument in his hands, Galileo turned it to the heavens and proved that the Copernican heliocentric theory was indeed true.

Here's a brief summary of his discoveries that led to that proof:

➤ Galileo first turned his attention to the Moon and discovered that it had many features that the Earth did. It wasn't the silvery, polished pearl that everyone thought it was. It had mountains and valleys, with some peaks as high as four miles. No real proof here, just a great discovery.

➤ He discovered Earthshine, the phenomenon that reflects light off the Earth on to the Moon. This led to the discovery that planets shine from reflected light from the Sun, and not because of their internal light, which was the widely held belief. Again this was no proof, but changed some basic beliefs about celestial bodies.

➤ After observing Jupiter, he found that it had four moons (there are actually twelve) revolving around it. The Earth was no longer the only planet that had moons, therefore less unique.

➤ The Sun, he discovered, had sunspots. By following the movement of these spots, he realized that the Sun turned on an axis. If this were also true for the Earth, the rotation on its axis would give the appearance of the Sun moving across the sky. So maybe the Sun is still and the Earth is moving. Now we're getting somewhere.

➤ Here's the real proof. Galileo found that when observing Venus, it went through phases just like the Moon. The way that light reflected off the planet changed its apparent shape; this was possible only if Venus orbited the Sun and not the Earth. Bingo!

Without the aid of the telescope, Galileo would never have been able to discover all of the things in the preceding list. Here is one of the first examples where technology was crucial in proving something was other than what it was thought to be. This ability to enhance our five senses and extend their abilities to perceive the world would lead to more and more inventions. Scientific instruments would become the hallmark of the scientific method. The study of cosmology from this time on would use instruments as an inherent part of its methodology.

Cosmonotes

Today we know that the universe is infinite, a concept that's not easy for us to really comprehend, since we have finite minds. But for the people of the middle ages and the Renaissance, the universe was considered finite. When people looked up at the stars in the sky, they believed that the distance to the stars was all the same and fixed. It was as if we were enclosed in a big shell. Heaven was on the other side of the stars. The idea of it going on forever wouldn't have entered their minds.

Kepler's Contributions

The last person that we'll look at in this chapter is Johannes Kepler (1571–1630). He was responsible for the discovery of a number of important concepts that added to the new Copernican cosmology. He was also a dyed-in-the-wool Pythagorean and neo-Platonist. So his theories would combine many of the ancient principles already covered along with a few more revolutionary insights. At the university where he studied mathematics, he became convinced of the new heliocentric model. But he needed to change and add a few more things to it before it could accurately account for all the observed celestial phenomenon. His greatest contributions can be described as follows:

➤ While the Copernican system correctly placed the Sun at the center of the known universe and Galileo's work proved its validity, neither of them considered the orbits of the planets to be other than perfect circles. This was still a Greek conception that hadn't changed. By observing the orbit of Mars in great detail and using the data collected by *Tycho Brahe,* he realized that the orbits of the planets were elliptical in shape rather than circular.

➤ Kepler explained the motion and distance of the planets from the Sun in terms of physical and mathematical principles called his three laws of motion. They can be summarized as follows:

1. All planets revolve around the Sun in elliptical orbits.

2. A planet covers equal areas in equal time as it moves through its orbit around the Sun.

3. The square of the time a planet takes for one complete revolution around the sun is proportional to the cube of the planet's distance from the Sun. (Can you pat your head and rub your stomach now?)

Universal Constants

Tycho Brahe (1546–1601) was an astronomer who developed his own theory of cosmology. His version included aspects of both Copernicus and Ptolemy. The Earth was still at the center with the Sun and Moon orbiting the Earth, but the other planets rotated around the Sun. He's not remembered for this theory as much as the incredible amount of extremely accurate measurements that he accumulated about the position of the Sun, Moon, and planets. Without this information, Kepler would probably not have been able to work out his three laws of planetary motion.

The displacement of earth from the center of the universe is probably one of the most significant changes in the history of cosmology. While arguments today focus more on the origin of the universe rather than earth's placement in it, in the sixteenth and seventeenth century the acceptance of the heliocentric theory didn't come easy. More than one person lost his life in support of it and many were imprisoned, but the damage had been done. The newly born scientific method would change the fundamental ideas of cosmology for centuries to come. In the next chapter, you will see how this method became the accepted way of understanding the world. And I think you'll be surprised to see that some of its strongest proponents still believed in the teachings of the ancient esoteric philosophy.

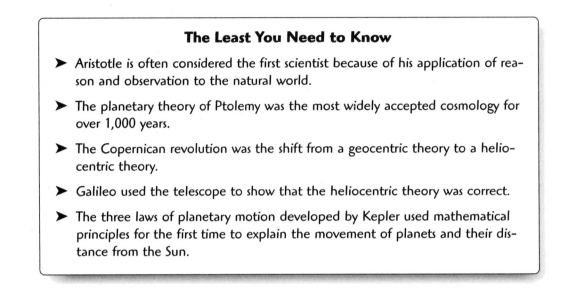

The Least You Need to Know

➤ Aristotle is often considered the first scientist because of his application of reason and observation to the natural world.

➤ The planetary theory of Ptolemy was the most widely accepted cosmology for over 1,000 years.

➤ The Copernican revolution was the shift from a geocentric theory to a heliocentric theory.

➤ Galileo used the telescope to show that the heliocentric theory was correct.

➤ The three laws of planetary motion developed by Kepler used mathematical principles for the first time to explain the movement of planets and their distance from the Sun.

Chapter 6

The Scientific Revolution

In This Chapter

➤ Things aren't always what they seem

➤ The different approaches to science

➤ Plato's theory of forms

➤ The impact of Francis Bacon

➤ The mind and body of Descartes

The foundations of cosmology changed considerably in the seventeenth century. At the heart of this change was a new way of understanding the natural world. The heliocentric description of the universe and the innovative methods used to prove it to be true eventually replaced the ancient cultural beliefs. But something was lost along the way. What that something was will be revealed as you explore this chapter.

We will also take a look at some of the philosophers and scientists who were responsible for this revolution in thinking. For they were the individuals who defined the brave new world and convinced everyone else that this was the best way to understand it. Little did they know the impact they would have on future generations. Let's take a look at another aspect of the Renaissance ideal to see how this all started.

Mindwarps

If you've ever tried dropping a feather and a rock at the same time, you would see the rock hit the ground before the feather. So how can it be true that objects do hit the ground at the same time? Well, if you put the feather and rock inside of a vacuum and drop them they will hit the bottom at the same time. You see, it's the friction of the air that keeps the feather up longer than the rock. Remove the air, and there's no friction.

What You See Isn't Necessarily What You Get

Do you remember the Renaissance ideal? Essentially it was the concept that knowledge of the world was based on human perception. But I need to add a corollary statement to that. Observation or perception of the world was important for knowledge. However, it was discovered that those observations might not be accurate. In other words, you need to question your perception. What happened to the perception that the earth was at the center of the universe? That, after all, was based on observation.

Another classic example came from our friend Aristotle. It seemed self-evident that heavier objects fell faster than lighter objects. This seemed so true that no one ever thought of testing it. It was only in the seventeenth century that Galileo tested the theory and found it to be false. He dropped a wooden ball and an iron ball from the Leaning Tower of Pisa and found that they both hit the ground at the same time.

Without testing, observations would either remain self-evident or a theory, maybe even part of dogma. This realization is at the core of the scientific method. It is what makes it such a powerful tool in which to study natural phenomenon. If you can set up an experiment to test a theory, the testing will show the theory to be true or false.

The questioning of perceptions gave birth to the *scientific revolution*. The idea of questioning our perceptions also led to inquiry in other areas. People began to question the doctrines of the church, the authority of kings and queens, and various beliefs about life and the world. Once it started, it couldn't be stopped.

Universal Constants

The **scientific revolution** describes a period of time from the middle of the sixteenth to the end of the eighteenth century. The basis for the word science comes from the Latin word *sciens*, which means "to know or discern." Added to that is the Old English word, *sceadan*, which means "to separate out." And that is exactly what science does. It seeks to know or discern the true nature of physical phenomenon by separating it out and identifying the principles under which it operates. This was a revolutionary way of studying the natural world, and it became the accepted method that is still used today.

Systems of Science

As with any significant change in a society, many different areas within the culture are affected, many of which can contribute to its further development. For example, look at how the computer has changed and impacted our lives in the last 25 years. Is there any area of our culture that has not been affected by it? It has become a global phenomenon, moving beyond the boundaries of any one culture.

The scientific revolution was no different. It would have been impossible for it to progress into the methodology it did without changes in other areas of thought taking place as well. One of these areas was philosophy, and the other was mathematics. Both of these are so intimately connected to the rise of scientific thinking that we need to look at how these two affected it.

Let's first formulate the scientific revolution in terms of ideas that are both philosophical and mathematical:

➤ It is important to understand how the world works.

➤ In order to do that, you have to examine the world itself rather than read Aristotle or consult scripture.

➤ A productive way to examine the world is through experimentation. Francis Bacon, whom you will learn about shortly, expressed this idea the clearest.

➤ The world is a *mechanical system* that can be described mathematically. This idea was introduced and clearly defined by René Descartes, whom you will also learn about shortly.

Universal Constants

A **mechanical system** can best be defined by using a model. Probably one of the best examples is a mechanical clock. If you look inside you can see an incredible interconnected system of gears, springs, and other moving parts that all make the whole thing work. Every piece is in one way or another affected by another piece, making the entire clock a contained system. One of the main qualities that are inherent in a system like this is predictability. If you know the position of any piece, such as a gear, for example, you can predict how its movement will affect the movement of another piece. This is, of course, the basis for a clock keeping accurate time. But it is also a property that will be used to define the mechanical nature of the universe.

Three Different Styles

There were three approaches to developing the proper scientific method. The first approach, which included the philosophers and scientists who felt that Aristotle's method worked best, preferred to analyze the nature of things. They used little mathematics and few experiments and sought to construct their system by logical arguments leading from a few basic premises. Their goal was to explain why things happen rather than to describe how they happened.

Cosmonotes

One of the goals in cosmology is to explain why things happen. This has fallen to the responsibility of only philosophers and theologians these days. The proper scientific method—that was eventually to develop—could explain how things happened, but not why. This is part of what was lost along the way. The earlier cosmological systems that were a part of Hermeticism and other branches of esoteric philosophy had always included why things happened the way they did.

The second system, which was led by men such as Francis Bacon and Tycho Brahe favored the *inductive* method. They felt that scientists should gather all the data possible through experiment and observation. Once collected, this data would point to the correct conclusion. If you remember, Tycho Brahe made an incredible amount of observations on the motions of the planets. His plotting of the periodic changes in the location of the celestial bodies led him to believe that Mercury and Venus revolved around the Sun, but that the Sun and the other planets orbited in turn around the Earth. He never reduced his system to a mathematical statement, but it did follow the observed facts better than the Copernican system.

The third system that was advocated at this time was the mathematical, *deductive* approach. Can you guess which mathematician influenced the scientists in this group? I'll give you a hint. It has to do with music. That's right, Pythagoras!

Many of the Renaissance humanists preferred Plato to Aristotle. And as you know, Plato was strongly influenced by Pythagoras and believed that all of the important elements of the universe were subject to mathematical demonstration. So he depicted nature in terms of straight lines, circles, triangles, and other geometric shapes that were more perfect than the objects actually observed. This was the deductive process of generalizing from basic principles, such as a geometric shape to a specific individual object.

It's All in the Mind

Plato's theory of forms, or theory of ideas as it is sometimes called, states that what is truly real is not the objects we encounter through our senses, but rather forms, which can be only grasped by the mind. An example would be the form circle. You never encounter a perfect circle in the real world. What you encounter are objects that have the quality of circularity. Sure you can draw a circle with a compass, but this is still an object that contains the quality or concept of circularity. The same is true of every object in existence. If I fill a room with a hundred different types of tables and I ask you to pick the one true table from all of them, which one would you pick? They're all tables that manifest as specific individual objects based on the general form—table.

Universal Constants

The **inductive** method utilized a way of thinking called inductive reasoning, in which you move from particular facts or individual cases to a general conclusion. When you see a particular thing occur repeatedly, you can make a general statement about the conditions in which it would happen again.

The **deductive** method is just the opposite of the inductive method and is based on a form of thinking called deductive reasoning. In this process you move from a known principle to an unknown, or from a general principle to specific case.

Under Plato's influence this approach to science became more mathematical than experimental. Among the chief supporters of the mathematical deductive approach of Plato and Pythagoras were Copernicus and Kepler. Galileo was also a great mathematician, but he chose *Archimedes's* teachings because he had applied mathematics to practical problems in physics. It was a method that Galileo would adopt and utilize to define two of the most fundamental qualities of the universe, mass and motion.

Here is a quote of Galileo's that expresses his thoughts on the role of mathematics in defining the structure of the universe:

> Philosophy is written in the great book which never lies before our eyes—I mean the universe—but we cannot understand it if we do not first learn the language and grasp the symbols, in which it is written. The book is written in the mathematical language, and the symbols are triangles, circles, and other geometric figures, without whose help it is impossible to comprehend a single word of it; without which one wanders in vain through a dark labyrinth.

And this brings us back to the two methods that are the main themes of the study of cosmology. Need I remind you what they are? Okay, one more time. The two methods are numerical symbols and observation. As you can see, observation has become a crucial part of the scientific method and mathematics has become the language in which science expresses itself. But there's more to this story. We need to look at a few of these philosophers and mathematicians to continue our exploration of how the scientific revolution materialized.

Universal Constants

Archimedes (287–212 B.C.E.) was a Greek mathematician, physicist, and inventor. He was probably one of the most brilliant of all the Greek mathematicians. His contributions include calculating the value of pi (the ratio of the circumference of a circle to its diameter), devising a mathematical exponential system to express extremely large numbers (he said he could numerically represent the grains of sand that would be needed to fill the universe), and inventing the screw. His most famous quote is "Give me a lever and a place to stand, and I will move the world."

Pass the Bacon, Francis

In the previous discussion on the second approach to developing the proper scientific method, I mentioned that one of the main adherents of this style was Francis Bacon (1551–1626). He is considered by many historians of science to be one of the main pillars that supported the new methods of science. Bacon saw himself as the inventor of a method that would kindle a light in nature, "a light that would eventually disclose and bring into sight all that is most hidden and secret in the universe." His method involved the collection of data, the carrying out of experiments, and organized observation to unlock the secrets of nature.

His three most well known works are: *Novum Organum* (or *true suggestions for the interpretation of nature*), *The Advancement of Learning,* and *The New Atlantis.* He used *analogies* a lot to get his ideas across. For example, the title page of the *Novum Organum* has an image of a ship passing through the *Pillars of Hercules,* which symbolized for the ancients the limits of humanity's possible explorations. The image represents the analogy between the great voyages of discovery and the explorations leading to the advancement of learning. The image also suggests that in using Bacon's new method, the boundaries of ancient learning will be passed.

Through all of his writings, Bacon believed that science needed to be practical and useful. He predicted many achievements of science and always had his eye on how it could benefit people's lives. In the end, he saw science as the increase of humanity's control over nature.

Universal Constants

An **analogy** is the abstract comparison of two dissimilar things. For example, you could say, "Your eyes remind me of two beautiful pearls." You're comparing two very different things, eyes and pearls, but showing their similarity by comparing qualities that both eyes and pearls have.

Descartes's Dualism

Modern philosophy began with René Descartes (1596–1650), mathematician, scientist, and philosopher. His importance to Western intellectual history can't be overestimated. He made important contributions to physiology, psychology, optics, and especially mathematics. He originated the *Cartesian Coordinate System* and most important of all, *analytical geometry.*

Universal Constants

The **Pillars of Hercules** refer to the two headlands on either side of the Strait of Gibraltar. This was the entrance and exit from the Mediterranean Sea to the Atlantic Ocean.

Universal Constants

The **Cartesian Coordinate System** allows you to represent mathematical formulas as geometric shapes: lines, curves, circles, etc. If you have studied algebra, you'll be familiar with plotting x and y coordinates on a graph to represent linear equations.

Analytical geometry is a combination of algebra and geometry. The position of any point in three dimensions (width, height, and depth, or on a graph as x, y, and z) can be indicated by algebraic symbols, and solutions are found by algebraic analysis. It was a necessary development in mathematics out of which would come calculus.

Descartes was a religious person, but he also believed there are important truths that can't be ascertained through the authority of the Church. These include those truths that pertain to the ultimate nature of existing things. But if that's the case, he thought, what is the criterion of truth and knowledge in such matters? What is to be the decisive factor by which you could separate certain knowledge about matters of fact from things based on just belief?

Descartes was intrigued with skeptical questions as to the possibility of knowledge, but he was not a skeptic. His interest in mathematics strongly affected his philosophical reflections, and it was his lifelong intention to formulate a unified science of nature that was as intrinsically certain as mathematics. But he did employ skepticism as a method of achieving certainty. (I'll make believe I'm Descartes so you can follow his train of thought.) His idea was simple enough: I will doubt everything that can possibly be doubted, and if anything is left, then it will be absolutely certain. Then I will consider what it is about this certainty (if there is one) that places it beyond doubt, and that will provide me with a criterion of truth and knowledge, which I can use to compare all other purported truths to see if they, too, are beyond doubt.

"I Think, Therefore I Am"

To doubt every proposition that he could come up with, Descartes employed two rather bizarre conjectures—the dream conjecture and the evil demon conjecture. "For all I know," Descartes said, "I might have been dreaming my entire life"—that is Descartes's dream conjecture. "And further," he said, "for all I know some malevolent demon devotes himself to deceiving me at every turn so that I regard as true and certain propositions that are in fact false"—his evil demon conjecture.

Mindwarps

The dream conjecture and the evil demon conjecture are totally bizarre. And that's the point. He knew how strange and wacky these two conjectures sounded. What Descartes was looking for was a measure of certainty that escaped even the most incredible and ridiculous possibilities of falsehood. In China the great sage Chuang Tzu echoed a similar dream conjecture. Upon waking from a dream in which he was a butterfly he asked himself, "Am I a man dreaming that I was a butterfly, or am I really a butterfly dreaming that I am a man?" (And he wasn't smoking opium either.)

What Descartes discovered when he considered everything he thought he knew in reference to one or the other of these two bizarre possibilities was that he could absolutely doubt everything except one indisputable truth: "I think, therefore I am." What this meant is that any attempt to doubt your existence as a thinking being is impossible because to doubt is to think and to exist. Try for a moment to doubt your own existence and you see what Descartes meant. The self that doubts its own existence must surely exist to be able to doubt it in the first place.

What this all means is that Descartes discovered in the certainty of his own existence an essential characteristic of certain truth: anything that was as clear and distinct as his own existence would also have to be certain. And the bottom line to all of this was that whatever is perceived clearly and distinctly is certain. Does this sound at all familiar to you? Can you see the connection to the Renaissance ideal? Knowledge is based on perception and this perception must be tested to make sure it is clear and distinct. That idea is the basis for the scientific method, and that is one of the reasons why Descartes played such an important role in its establishment.

The Mind/Body Problem

Descartes then had a means to go about analyzing the world around him. In that process he discovered that reality seemed to have a dual nature. On the one hand there are objects outside of us that occupy space, and then there is the mind inside of us that perceives these things. The objects in the outside world require nothing other than themselves to exist, so Descartes thought that mind and matter were totally independent of each other.

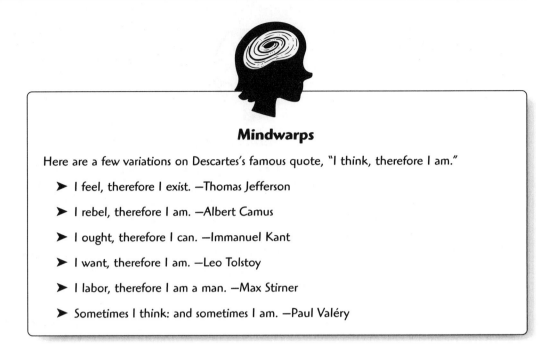

Mindwarps

Here are a few variations on Descartes's famous quote, "I think, therefore I am."

➤ I feel, therefore I exist. —Thomas Jefferson

➤ I rebel, therefore I am. —Albert Camus

➤ I ought, therefore I can. —Immanuel Kant

➤ I want, therefore I am. —Leo Tolstoy

➤ I labor, therefore I am a man. —Max Stirner

➤ Sometimes I think: and sometimes I am. —Paul Valéry

The problem that he ran into was that, in people, the mind and the material body interact; the motion of the body is sometimes affected by the mind, and the thoughts of the mind can be influenced by outside physical events. He ascertained that material things, including one's own body, are completely subject to physical laws, but the immaterial mind can move one's body.

Cosmonotes

Much of the material that I'm covering here is a simplified version of many of the insights, theories, and arguments that Descartes and the seventeenth-century philosophers developed. If you want to delve into Descartes's conjectures and the ideas of other philosophers and scientists, I would highly recommend checking out books on the history and philosophy of science (for example, Will Durants's *The Story of Philosophy* or Frederick Copleston's *History of Philosophy* series).

He found it difficult to understand just how something immaterial as the mind could move something material, like the body. And this further implied for him that if the immaterial mind can do this, then our bodies were evidently not completely subject to physical laws.

If this all sounds a bit confusing, let me shed some light onto what this all has to do with cosmology and why it's important to look at. Descartes's overall approach to studying the physical world was twofold:

➤ As a mathematician and scientist he wanted to develop a system of inquiry within the scientific method that was as certain for him as his realization that I discussed earlier. And for him this system would be based on the certain truths found in mathematics that would explain the operation of the universe in clear, precise, mechanical terms.

➤ His other approach had to do with how the world was comprehended by the mind of the observer. Instead of asking, "What is the basic stuff of the universe?" Or "Under what laws does the universe operate?" Descartes asked, "What do I know is the basic stuff?" and "Of what can I be certain regarding the laws of the universe?" In other words, Descartes was questioning the ability of the mind doing the perceiving to correctly interpret what was presented to it.

Lost Along the Way

The scientific method in the twentieth century continually sought to construct experiments carefully enough so that the mind doing the observation did not have to be accounted for in the experiment. For the most part, almost all of the scientists prior to the last century didn't even think about the mind as an issue. Descartes clearly was one of the first thinkers to do so, but because it was difficult to ascertain exactly what the mind's role was in observing an experiment, these ideas disappeared from the scientific method and were relegated to realm of philosophy.

But as we will see in later chapters, any observations about the natural world or the universe have to be, in part, observations of the mind of the observer as well. Other philosophers besides Descartes felt that this was a sufficiently important question to devote considerable amount of time and energy to addressing this issue. We will take a look at some of these individuals later on as well.

At least for now, you should have a clearer understanding of how the scientific revolution of the sixteenth century began. In the next chapter, we will look at a few more very important people who were instrumental in making science the accepted method of studying the universe. But the basis of their contributions will be built on aspects of ancient wisdom as much as modern science.

Before I close this chapter I wanted to ask you if you discovered the something that was lost along the way that I alluded to back in the beginning of this chapter. I made a brief reference to it in one of the sidebars. Part of what science lost was the ability to provide answers to why the world operated the way it did. Its strength lies in its capacity to explain how things work, not why they work that way. The other part that was lost was the organic quality of the universe. It was the sense of it being alive in a certain way, not necessarily in terms of divine action, but that sense of being connected to it and that part of it was disappearing. The world outside of us became sterile. Matter was dead and dumb. The universe and the world became an object to be categorized, dissected, and analyzed. Reason and rational thought replaced intuition and organic wholeness. The ultimate goal became the control of nature.

The Least You Need to Know

➤ The Scientific Revolution occurred as a direct result of people questioning their perceptions about the world.

➤ Because of Plato's influence science became mathematical as well as experimental.

➤ Francis Bacon believed that the goals of science should be for the comfort and benefit of humanity, with the ultimate goal being the control of nature.

➤ René Descartes invented a new form of mathematics called analytical geometry, which combined algebra and geometry.

➤ Descartes felt that it was more important to question our perception of nature than nature itself.

➤ Science describes how the universe operates, not why.

Chapter 7

Cosmology in the Age of Enlightenment

The period from 1650 to 1800 is often called the Age of Reason or the Age of Enlightenment. The name implies a predominant way in which people looked at the world. Seventeenth- and eighteenth-century thought held that people were rational beings in a universe governed by some systematic natural law. Some believed that law to be an extension of God's law. Others felt that natural law stood by itself.

Natural law was extended to include international law, and accords were formulated in which sovereign nations, bound by no higher authority, could work together for the common good. The fever of rationality spread throughout European thought and brought with it great hope and some distrust.

What Is Enlightenment?

When you hear or see the word faith, it's most often associated with religion. But you can have faith in other things as well. Faith in yourself, your friends, sometimes even your government. In the Age of *Enlightenment* faith was placed in science, in human rights arising from the natural law, in human reason, and in progress. The idea of progress was based on the assumption that the conditions of life could only improve with time and that each generation made life even better for those following.

Universal Constants

The **Enlightenment** generally refers to eighteenth-century Europe when rational thought and reason dominated social and political thought. At the root of the word is the term enlighten—to reveal truths, to free oneself from ignorance, prejudice, and superstition. And that's how people felt—that science and rational thought were leading humanity out of the darkness and into the light.

Enlightenment, reason, and progress are secular ideas, and the age became increasingly secular. Politics and business superseded religion, wrestling leadership away from the Church, of whatever denomination. Toleration increased and persecution and the imposition of corporal punishment for religious, political, or criminal offenses became less common as the era progressed.

However, not everyone jumped on the Enlightenment bandwagon. Some people felt that while the new scientific method was great, it couldn't possibly account for all phenomena in the world. There were some that distrusted the premises of science and were uncertain of its impact on society. The French philosopher Blaise Pascal gave perhaps the greatest expression to the uncertainties generated by the scientific revolution. He wrote:

> For, after all, what is man in nature? A nothing in comparison with the infinite, an absolute in comparison with nothing, a central point between nothing and all. Infinitely far from understanding these extremes, the end of things and their beginning are hopelessly hidden from him in an impenetrable secret. He is equally incapable of seeing the nothingness from which he came, and the infinite in which he is engulfed. What else then will he perceive but some appearance of the middle of things, in an eternal despair of knowing either their principle or their purpose. All things emerge from nothing and are borne onwards to infinity. Who can follow this marvelous process? The Author of these wonders understands them. None but he can.

One of the cornerstones of the Enlightenment that had developed out of the scientific method was the idea of natural law. It, like many other suppositions of this age, was a secular or worldly idea not based on religious traditions. Natural law, or simply the laws of nature were not only the physical laws that governed the many aspects of nature, but was also a principle of rationality that infused the universe, to which human behavior ought to comply. It became part of the political arena and its concepts formed the basis for more than one social reform. You can find it as one of the fundamental ideas in our Declaration of Independence.

Whose Science Is Whose?

It's a good idea to provide a general description of some of the main *paradigms* or worldviews that people held at this time. From this we can explore the current state of cosmology and look at some particular aspects of it. Here's a breakdown of some of the main belief systems that manifested at this time (the following are numbered only so that you can refer to them):

1. Science was based on mathematical formulas, experimentation and observation. Rational thought was the superior way of analyzing events going on in the world. Science led to progress and dominion over nature and was the only reliable method for understanding the structure of nature. Religious dogma and Church authority were not sources to be considered because they sought to control people's minds and free thinking.

Universal Constants

A **paradigm** is a pattern or model, and more specifically, a worldview. It is often used to describe collective and individual perspectives of the world. For example, you can have a scientific paradigm, a Christian paradigm, or a democratic paradigm. These collective worldviews will share general beliefs and principles. On an individual level, our paradigms reflect our values, beliefs, knowledge of the world, and life experiences. These paradigms are in a constant state of flux as new knowledge and experience replace, enhance, or update their contents.

2. Religion was the only true source with which to understand the world. Science could be useful, but shouldn't be the preferred method. It had no spiritual center and was atheistic at its core. It couldn't account for the creation of the universe and served only humanity and not God. Some aspects of it should even be distrusted. It sought to put humanity above God.

3. Science and religion could work together to explain the nature of the universe. One could still be religious and embrace the scientific method. God gave human beings minds with which to understand His creation. The scientific method was a means by which people could fathom the inner workings of the divine design.

4. While science incorporates a specific methodology with which to study the universe, the goals of science are formulated on the principles found in esotericism. The scientific method is a means to show that the universal truths found in esoteric philosophy are the fundamental laws that govern the structure of the universe.

Black Holes

Ideas are very powerful tools of change. In them is locked the potential to bring about increased understanding for the benefit of society. But they can also be used to justify the darker actions of humanity. While the ability to question religious authority and the rule of monarchs freed people from centuries of control, taken to an extreme it brought about years of religious wars and turned the French Revolution into "The Reign of Terror." Ideas taken to any extreme are dangerous, for then people, instead of using ideas, are used by them. It doesn't matter if it's in the realm of science, religion, philosophy, or politics, ideas are neither good nor bad, it's what people do with them that matters.

The four categories outlined in the preceding list are very general in nature. There are others that could be included that would contain various aspects of other collective paradigms; some that are more political in nature, militaristic, or even economic. During the Enlightenment, as during the Renaissance, tremendous change was taking place. There were the revolutions against the long established monarchies of Europe.

The seeds of the Industrial Revolution were also being sown. War and conquest were occurring in Europe with the likes of Napoleon, and in the colonial territories. Technological inventions occurred at an astonishing rate, forever changing the makeup of society and how people lived their lives. But since we're focusing on cosmological paradigms, these four are more appropriate for our discussion. Let's take a closer look at each one of these and meet some of the people and ideas associated with them.

Science Is King

At the end of the seventeenth and early part of the eighteenth century, the most significant figure in the field of science was Isaac Newton. His discoveries affected science, especially physics and cosmology, for the next two centuries. He still plays an important role today, but to a lesser degree. Normally he would be associated with number three in the preceding list, but surprisingly he belongs with number four. You'll find out why by the end of the chapter.

Cosmonotes

In the seventeenth century, scientists realized that a standardized unit of measure was needed. They had to define new units of measurement previously unknown: measurements of time, temperature, pressure, weight, length, etc. New ideas gradually gained ground, but only when there was a unified system of measurement could such a basic concept become an effective instrument. In 1790, the decimal metric system was created.

There are remarkably few well-known scientists who would actually belong to group number one. All the ones we have covered so far—Galileo, Kepler, Descartes, and Bacon—while championing the scientific method, were still religious at heart. There were others who were strictly rationalistic and even atheistic, who did belong to the first group, but they never discovered or published anything that would be considered important to science. Most of the writings we do find related to the first group came from people who weren't even scientists, just people who adopted the ideas behind science. Some of these were philosophers, others were historians, and some literary writers.

While there definitely were individuals who used the tenets of science to expound their philosophical theories or apply them for political ends, there were others who embodied the ideals of science and put them to practical use. After all, the Enlightenment was essentially a movement of the middle class. With economic prosperity on the rise, and the Industrial Revolution about to take hold, it was the middle class of society that sought to transform itself into the ruling majority. Technological progress was the key and science found its home in the inventions of the average person.

The Word of God

In direct opposition to group number one in the preceding list, was the second group. This paradigm was purely religious and just like the first saw its beliefs as the only true source of knowledge about the world. The Catholic Church was still reeling from the Protestant Reformation that had occurred in the first half of the sixteenth century and saw its position as the spiritual authority undermined even more by the advancement of science.

Natural Philosophy was what science was called from the time of the ancient Greeks through the nineteenth century, so scientists during the Enlightenment still referred to themselves as natural philosophers. Natural philosophy and the Church really didn't have problems with each other until the Copernican Revolution. Until then, most of what was found in the Bible agreed with the accepted cosmology of Aristotle, Plato, and Ptolemy.

As previously discussed, the early scientists that questioned the teachings of the Church got into serious problems. Giordano Bruno was burned at the stake; and Galileo was imprisoned, tortured, and lived out his life under house arrest. But the scientists weren't the only ones that suffered at the hands of the Inquisition. With the advent of the Protestant Reformation headed by individuals such as Martin Luther, Erasmus, Ulrich Zwingli, and John Calvin, Europe plunged into religious wars that lasted from the sixteenth century to the end of the seventeenth.

Black Holes

The practitioners of the pre-Christian religion of Europe also suffered at the hands of the Church. Known today as Wicca, the followers of this ancient nature-based religion were persecuted as followers of the devil because of ignorance of their true beliefs. Between 1100 and 1700 C.E. various sources estimate that 300,000 to 800,000 people, mostly women, were executed or tortured for practicing witchcraft.

Many people, Protestant and Catholic alike, considered the Bible to be the literal word of God. For them and the Church it was an accurate history of the creation of humanity and the universe. There was no need to look any farther and to do so would be to doubt God's plan and authority. Science and the philosophies that went with it questioned the authority of Church, the Bible, and even God.

To show how accurate of a history the Bible was, a bishop in the seventeenth century went back through it, putting together a chronological table of all of the generations from Adam through Jesus and figured out that the world was created in 4004 B.C.E., at 2:30 P.M. on Sunday October 23rd. That makes the universe about six thousand years old. Hmmm, I guess he was a little off.

Half and Half

In one corner we have the scientific skeptics, reveling in their freedom and ability to question everything and show by the brilliance of their logic and methodology the fallibility of anything that crosses their path. In the other corner are the true believers, who know beyond a shadow of a doubt that their religion is the one and only true source of knowledge. In the middle between these two are found the humanists. You met their forerunners back in Chapter 4, "Transitions in Thinking."

Okay, quick memory check, what is humanism? (Game show music playing in the background ....) Yes, that is correct! Humanism was a philosophy that allowed humanity to fulfill itself apart from religious doctrine. Humanity had the potential to find earthly fulfillment through the realization of its own inherent goodness.

Mindwarps

One of the greatest and most influential figures of the Enlightenment was Voltaire. He wrote plays for the theatre as well as philosophical treatises. His international fame gave him widespread opportunity to speak out against religious intolerance and to support material prosperity, respect for all humans, and the abolition of torture and useless punishment. He was known in all of the royal courts of Europe and England and embroiled himself in more than one political cause, usually for the benefit of the lower classes and serfs. He embodied the essence of humanism.

By the end of the eighteenth century the humanistic movement had tried to incorporate the best parts of what the other two groups had to offer. In this group you would find religious people who adopted the scientific method. Rational thought was the focal point of their thinking, and they used it to support their religious views as well as their scientific views. This was an eclectic group that included bishops, a few popes, some well-known literary writers, and a number of scientists and philosophers. Out of this third group would come the individuals who would have the most positive impact on the greatest number of people.

Deism was the main religious belief system of many in the third group, along with a nonacceptance of the concept of original sin (the first sin committed by Adam, which "stained" all of humanity thereafter). I think it's a good idea to define here some of the various religious and nonreligious beliefs systems that were prominent during this time.

➤ **Theism** The belief in one God who is the creator and ruler of the universe and is known through revelation. He continues to actively participate in the affairs of humanity.

➤ **Atheism** The belief that there is no God, or is the denial of God's existence.

➤ **Deism** The belief in God on strictly rational grounds without reliance on authority or revelation. God created the world and all its natural laws, but takes no further part in its functioning.

➤ **Agnostic** The belief that the human mind is incapable of knowing whether or not there is a God or an ultimate cause. It can also pertain to people who haven't decided one way or another and are waiting for more information before making up their minds.

Since science had no theories yet for the origin of the universe, accepting the creation of the universe as a divine act seemed to make perfect sense. But that's about as far as many of the humanists wanted religion to go. It could supply a moral code as to how you could live your life, but as far as other dogmatic doctrines went, they bailed on most of those.

Discussion in this last section will deal with the role that esoteric philosophy played in the lives of a couple imminent scientists. The focus will be especially on alchemy. In recent years there has been scrupulous research done regarding the impact that this ancient art had on the development of certain doctrines in science. Science historians have often sought to downplay the role that alchemy had in the scientific revolution. In general, science has taken a simplistic and somewhat arrogant stance against alchemical practices. In this section I'd like to bring to the readers' attention, in the light of more recent evidence, that alchemy was much more influential than originally thought.

The Science of Alchemy

Our last group has the smallest number of people in it of the four. By the end of the eighteenth century, most of the teachings of Hermeticism had either fallen to the wayside or had been incorporated into larger bodies of knowledge, losing its distinctiveness and origination. There is one aspect of it, however, that continued to strive behind closed doors: the practice of alchemy. And alchemy is really what lies at the heart of Hermeticism and much of esoteric philosophy. Surprisingly, cosmology and alchemy have probably more in common than not. The rest of this chapter will focus on some of those similarities and others will pop-up in chapters down the road. For now, let's look at what alchemy is.

Lead into Gold

The word alchemy comes from the Arabic word *Al-kimiya*, which is derived from the Egyptian word *Keme*, meaning "black earth," the name given to Egypt based on the annual flooding of the Nile and its deposits of rich dark silt. Since it is an essential part of the Hermetic teachings, it would also have originated in Egypt.

In general, science does owe a debt of gratitude to the alchemists, for many of their discoveries and laboratory techniques gave birth to chemistry. And in a nutshell, that's how alchemy is viewed by science, as the forerunner of chemistry and nothing more.

Cosmonotes

One of the goals of alchemy has been the search for the philosopher's stone—a substance that *is* said to have the capacity for turning metals such as lead, tin, copper, and other base metals into gold. Critics of the alchemical tradition sight the fact that no one, to their knowledge, has ever found this substance. Consequently the search must be based on something that in reality is unattainable. However, practical alchemists were well aware of the fact that if they succeeded in making gold artificially, their lives would be in grave danger from the avaricious princes and other evilly disposed people. There is some information on alchemists who did claim to have produced the philosopher's stone. One of the most famous, Nicolas Flamel, donated a fortune to various charities during his life, while living an obscure and very plain existence.

Alchemy is the art and science of transmutation, but the transmutation of what? Lead into gold. And that is where much of the controversy, lack of understanding and short-sightedness comes from. Alchemy like other areas of study has two sides to it, the exoteric and esoteric. The exoteric is what historians have spent much of their time discussing.

The esoteric side is what the true alchemist was interested in. The transmutation of lead into gold is only an outward symbol for an inner transfiguration. Alchemical symbolism is very unusual and sometimes outright bizarre. It can take years, even a lifetime of study to decipher the symbols, and critics again dismiss the symbolism because they don't know the hidden language to be found in them.

To quote present day alchemist Jean Dubois:

> Alchemy is the Science of Life, of Consciousness. The alchemist knows that there is a very solid link between matter, life and consciousness … Alchemy deals with the inner dimensions of life, the meaning of life, and the relationship between spirit and matter. Work in the laboratory is where the outward sign of an inner transformation is reflected. You will transmute nothing if you have not transmuted yourself first.

And on a side note, Monsieur Dubious, who is the founder and first president of the French alchemical organization *The Philosophers of Nature,* is also an electrical engineer with a large firm in France, and has worked in the field of nuclear physics with more than one Nobel Prize winner.

In the twentieth century, the well-known Swiss psychologist Carl Jung became a proponent of alchemy. He noticed that in dreams that both he and several of his clients had there were resemblances to many alchemical symbols representing stages of self-development. However, for Jung, the entire alchemical work was viewed strictly from a psychoanalytical perspective.

Cosmonotes

If you're interested in reading more about Robert Boyle and his interest in alchemy, I highly recommend *The Aspiring Adept* by Lawrence M. Principe.

So this should give you a little background on some of the ideas associated with alchemy. Let's turn our attention to two very important scientists/alchemists who had a profound impact on the cosmology of the eighteenth century: Robert Boyle and Isaac Newton.

It All Boils Down to Boyle

Robert Boyle (1627–1691) is considered to be the father of modern chemistry. His contributions to the development of science include his air-pump experiments and the fundamental gas law that bears his name. His most famous work, *The Sceptical Chymist,* was always considered the text that broke away from the irrational and misguided practices of alchemy. But more recent scholarly analysis has shown that this work doesn't criticize alchemists, but instead "unphilosophical" pharmacists and textbook writers. In addition to this, his "lost" *Dialogue on the Transmutation of Metals,* was recently pieced together from fragments for the first time, revealing its alchemical nature.

Boyle was a lifelong advocate of the alchemical tradition. While still in support of the scientific method, he didn't accept its mechanistic view of the universe. A mechanical universe, functioning regularly by the operation of physical laws, no longer needs God to keep it running. Although this is a form of deism, Boyle was too religious at heart to remove God that far away from his creation. Alchemy, like much of esoteric philosophy, sees the world as organic, interactive, and fundamentally spiritual at its core. So Boyle's experiments and core philosophy were essentially alchemical, and his science was a direct result of this tradition.

Newton's Alchemical Apple

Isaac Newton (1642–1727) is best known as a founder of modern science. He was a great mathematician and physicist who invented *calculus,* found the law of universal gravitation, and satisfactorily explained the spectrum of colors in rainbows and prisms. It's true that he did all of these things, but he spent most of his time on alchemical and theological studies. Recent scholarship has shown, as with Boyle, that there was another side to the scientific genius.

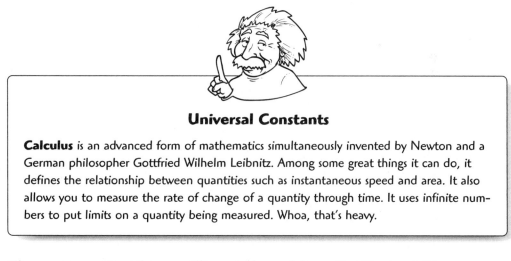

Universal Constants

Calculus is an advanced form of mathematics simultaneously invented by Newton and a German philosopher Gottfried Wilhelm Leibnitz. Among some great things it can do, it defines the relationship between quantities such as instantaneous speed and area. It also allows you to measure the rate of change of a quantity through time. It uses infinite numbers to put limits on a quantity being measured. Whoa, that's heavy.

There are approximately one million words on alchemy that Newton left in manuscript, and it's only been recently that they have been seriously studied. Included among the manuscripts is a lengthy commentary that he wrote on the *Emerald Tablet of Hermes.* The problems that he hoped to solve through his study of alchemy were very similar to Boyle's. It again goes back to the notion of the mechanical universe.

One of the main tenets of this philosophy was that events in the natural world were explained by matter and motion, by tiny particles of a passive, inert matter that transferred motion by pressure or impact with other particles. How can passive little billiard balls of matter cohere and stick together in organized forms? How can those tiny passive particles of a common universal matter combine to produce the immense variety of living forms found in the world?

Mindwarps

Although the discovery of the atom and molecules was still almost 200 years away, many scientists and philosophers believed that matter was made up of small particles called atoms. This was a notion handed down through the ages from the ancient Greeks who first came up with the idea and coined the term atom. The only thing that had changed was instead of matter being made up of various different kinds of atoms, it was thought that matter was made up of one kind of universal element. It was the search for this element that formed part of the alchemical quest.

Newton felt that the passive particles of matter couldn't organize themselves into forms without divine guidance. This divine guidance was believed by Newton to be some spirit of life that could guide the passive matter into the beautiful forms of plants, animals, and minerals that God had ordained. It was this spirit of life that Newton hoped to learn from alchemy.

Many alchemists, including Newton, made connections between events in the Bible and alchemical processes. The story of creation in Genesis was considered to be an allegorical story that veiled hidden alchemical messages. And themes of alchemical death and resurrection were symbolized in the death and resurrection of Jesus. All in all, the alchemical tradition was an important and vital part of the paradigms of many scientists of the eighteenth century.

In the next chapter, I'll cover the state of cosmology up to the beginning of the twentieth century and look at it from the point of view of the microcosm and the macrocosm. We'll even compare notes on creationism vs. evolution, still a hot topic today. So ring the bell, close the book, and quench the candle, and remember that just because you believe something is true doesn't mean it really is.

The Least You Need to Know

➤ The Age of Enlightenment put its faith in science, human rights arising from natural law, human reason, and progress.

➤ Deism, or the belief in God on rational grounds and not revelation or authority, was a popular paradigm during the Age of Enlightenment.

➤ Alchemy was an intrinsic part of some scientists' research because they were dissatisfied with the mechanistic view of the universe.

➤ Isaac Newton was a great scientist and physicist, but he spent much of his time studying alchemy and relating it to scripture.

Physics 101

The technological advancements of the eighteenth and nineteenth century allowed the study of cosmology to branch into two main areas of investigation. What the telescope did for examining the heavens, the microscope did for exploring the world beneath our feet. This was the beginning of what would also become the two most significant areas of study in cosmology today. Astronomy and astrophysics explore the depths of the universe with telescopes that would have been science fiction 200 years ago. The other branch of cosmology, which today falls under the category of particle physics, began with the discovery of atoms and molecules just less than 100 years ago.

The main focus of cosmology prior to the twentieth century was still on the skies. The thought of including the study of what the universe is made of fell more into the realm of chemistry and various categories of physics such as the study of heat, magnetism, and electricity, rather than cosmology. But all of the knowledge gained from these different areas came together in the twentieth century to bond the study of the microcosm to the study of the macrocosm.

Honey, I Shrunk the Universe!

The main goal of this chapter is to bring you up to speed on the state of cosmology just at the beginning of the twentieth century. We've come from ancient Meso-potamia to the start of Newtonian cosmology. That's a lot of cosmological history to cover, but from this point on we only have a short span of time to cover. The difficulty in explaining where cosmology was and where it was going at the end of the eighteenth century is because of all the diverse areas of science that went into it. In a little while I'll summarize for you some of the different areas of science that would be necessary for it to become the cosmology of the twentieth century. For now, let's look at the importance of the microscope.

When all you have to look at the world are your eyes, you're limited to understanding its structure, because you don't realize there's more to it than just what can be seen with the naked eye. We saw how the invention of the telescope enabled Galileo to see things in the heavens that no one had ever seen before. And that was just the beginning.

Mindwarps

Before the invention of the microscope, doctors and scientists could only observe what they could see with the naked eye. Early microscopes used drops of water captured in a small hole to create magnification. The first microscope, like the first telescope, used ground glass as a magnifying lens—just in reverse. Instead of magnifying what was far away, it magnified what was up close. Later on, a reflecting mirror was used to reflect light back to a focal point instead of refracting it, which the optical lens did. Eventually in the twentieth century the electron microscope was invented that used a beam of electrons to illuminate objects instead of light, and an atom was seen for the very first time.

The same sort of thing happened at the end of the seventeenth century when Anton Van Leewenhoek invented one of the first microscopes. All of a sudden a whole new world opened up before people's eyes. The power of magnification revealed a world of tiny organisms not normally visible to the naked eye. With more improvements it helped to identify and observe the cells of plants and the tiny blood vessels in our bodies called capillaries, which link arteries to veins. This provided enough evidence for the development of theories about the circulation of blood through the human body. The microscope would become the most important instrument in biology and medicine.

The telescope changed cosmology in a way that the microscope never would, but the important point about the microscope was that it made people aware of the fact that there was a universe to be explored in the opposite direction from the cosmos out there. A universe within a universe you could say. And this idea of exploring the *microcosm* eventually led to the quest for the smallest particle of matter that could exist. It's the same quest that is still going on today.

Now, it's not that no one ever thought about what the most basic constituent of matter was, or what the smallest particle might be. Remember the early Greek philosophers like Empedocles, who thought that everything was made up of four elements? (And I know you know what they are.) And some of the others who wondered about the one prime material that everything came from. It's just that for the first time, instead of just theorizing and thinking about it, the microscope gave proof that the world and maybe the entire universe was made up of smaller and smaller particles. They thought, "Maybe if we invented a microscope strong enough, we could see how small everything really is." They had the right idea. It would just take a while for it to happen.

Universal Constants

Microcosm simply means little world. It comes from the two Greek words, *mikros* meaning minute or tiny, and *cosmos,* which you already know means world or universe. It can refer to anything that is a small self-contained system. In nature you can see this in the ecosystem of a pond. It's a relative term, too. Our planet earth would be a microcosm compared to the rest of the solar system. And in esoteric philosophy, humanity is the symbol for the microcosm of the universe. All that is within the universe is also within us. As they say, we are such beings as stars are made of.

I'll come back to the examination of the microcosm in a few chapters. You'll see that the study of cosmology at the microcosmic level is where many of the hottest topics in science are being explored. But enough for now, I don't want to give too much away. Let's get back to the other fields of science that made twentieth-century cosmology possible.

Mindwarps

Around 400 B.C.E. there was a group of Greek philosophers called the Atomists. Actually there were only two, Leucippus and Democritus, but I guess that's enough to call them a group. The basic idea and use of the word atom is attributed to Leucippus, and the detailed working out of Atomism is considered the result of Democritus. Atomists held that all things are composed of physical atoms—tiny, imperceptible, indestructible, indivisible, eternal, and uncreated particles composed of the same matter but different in size, shape, and weight. By combining in various ways, atoms compose the objects of existence. Little did they know how close to reality they really were. It's no wonder that the term atom was used when they were finally discovered.

Can't Forget About Math

Of all the fields of study that can be found within the realm of physics there are four that were crucial to the later exploration of cosmology:

➤ Temperature, especially the understanding of heat and its generation

➤ Magnetism and its relationship to electricity

➤ Electricity and its components

➤ Optics, the study of light, its structure, and how it can be used

Chemistry is very important as well, but that's an area of study all unto its own. However, we do need some help from it too before the picture of twentieth-century cosmology can be painted.

The one thing that all of these areas of physics have in common is that they are all forms of energy. And energy, we now know, is what everything is made of. However, two hundred years ago no one knew this. The categories mentioned above were all separate, and the understanding of their true relationship would only slowly be revealed.

For a moment, I want to touch on our two original themes that I said were the most important ways we have developed to study the origin, structure, and evolution of the universe: the observation of the heavens and the use and application of numerical systems. That last method I'm going to replace with the term mathematics, because by now that's what it has evolved into.

The history and development of mathematics is as important as the study of physics to our understanding of cosmology. Without math, the ability to formulate and understand the four categories mentioned in the preceding list would never have occurred. However, I don't want to spend any more time on it other than to provide you with a short summary of its development. The order of its progression went something like the following:

➤ Numbers were first used as symbols to denote amounts and quantities of things. For certain cultures they also had a deeper significance as already explained in the chapters on the early creation stories as well as for the Pythagoreans and Plato.

➤ Geometry and trigonometry were both used by the Mesopotamians, Egyptians, and Greeks. Geometry uses numbers to define the properties of objects in space. Points, lines, surfaces, and solid shapes like cubes, pyramids, and others were measured, and their mutual relationships were defined. Trigonometry is a natural progression from geometry. It utilizes the knowledge about geometric shapes, mostly the right triangle, to find unknown angles and sides of a triangle. This is done by defining ratios between any side of a right triangle and one of the acute angles. (That's an angle less than 90 degrees.) It's important for surveying, navigation, building, and engineering.

➤ Algebra was the next development. Geometry and trigonometry were around for more than 2,000 years before this new form of mathematics was developed. It came to us from the Islamic countries during the Renaissance. They had worked it out in the eighth century along with the number 0. Algebra primarily deals with abstract thinking in which letters or other symbols are used to represent numbers. These are manipulated in various ways to find unknown quantities.

Mindwarps

Many ancient cultures didn't use numbers; they used letters. The Greek and Hebrew alphabets were used to represent numbers, similar to the Roman numerals, except each letter of the alphabet had a numerical equivalent. This means that the spelling of any name in these alphabets also had a total numerical value. This is the original source for numerology and the study of the Cabalah.

➤ Along with algebra came the mathematical symbols we use to represent numbers today. They are very close—almost identical—to Arabic numbers. Islam inherited them as well as the mathematical system of adding, subtracting, multiplying, and dividing from India. Have you ever tried to do math using Roman numerals? It's impossible!

➤ After algebra came analytical geometry, invented by whom? You know, "I think, therefore I am."—René Descartes. To refresh your memory of what this form of math was, go back and reread the definition of it in Chapter 6, "The Scientific Revolution."

➤ Then came calculus or, as Newton called it, fluxions. Thank God we changed the name. Anyway, doing calculus is much easier than trying to describe what it is. However, I think I gave you a pretty good definition in the last chapter. If you don't remember, go take a peek.

➤ And, of course, there are many more forms of mathematics that have developed since calculus, but there's no need to go into those. They deal with complex numbers, imaginary numbers, infinite number sets, set theory, Boolean algebra, differential equations … need I say more?

You're Getting Warmer

Okay, that's enough about math. Let's move on to the four categories of physics that were so crucial to the advancement of twentieth-century cosmology. The first one I will discuss is heat. In and of itself, the word means little other than the opposite of cold. It defines a relative range of temperature from warm to hot. In physics, heat is defined as a form of energy that does work.

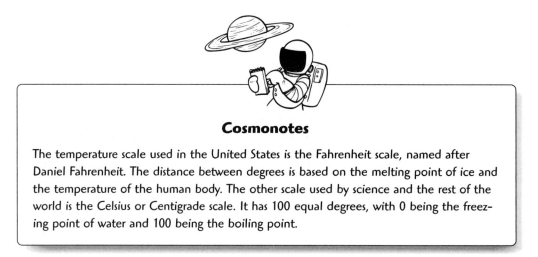

Cosmonotes

The temperature scale used in the United States is the Fahrenheit scale, named after Daniel Fahrenheit. The distance between degrees is based on the melting point of ice and the temperature of the human body. The other scale used by science and the rest of the world is the Celsius or Centigrade scale. It has 100 equal degrees, with 0 being the freezing point of water and 100 being the boiling point.

The study of heat really began when Galileo invented an instrument called a thermo-scope. There was no scale on it to read; the idea of degrees of temperature didn't come until later. But his sealed glass tube that was filled with colored water registered temperature changes in the surrounding environment, and the gas that was trapped inside the tube contracted or expanded, moving the column of colored water up or down. The expansion and contraction of sealed gases was the basis for the first thermometers.

Since the first thermometers used the expansion and contraction of gases to measure changes in temperature, the next logical step would be to study the nature of gases. Based on experiments done by our alchemical friend Robert Boyle, and later Jacque Charles (both have gas laws named after them, which is an honor I guess, but I think I'd like to be related to something other than gas, if you know what I mean), it was discovered that all gases expand and contract in the same way. They also found a direct correlation between temperature, pressure, and heat.

Nobody really knew what heat was. If you've ever looked at something very warm or hot, like a black asphalt road in the sun, or a fire against a light background, you may have seen wavy, colorless lines moving through the air. And that's what many people thought heat was: a colorless, invisible fluid that was able to permeate or penetrate whatever it came in contact with.

It wasn't until James Prescott Joule performed an ingenious experiment that the real breakthrough came in understanding what heat was. His experiment transformed mechanical *energy* into heat. The accompanying realization with this experiment was that heat is a form of energy that can do work. (Too bad you can't get it to go to work for you on the weekend.) And heat is related to the three primary states of matter—gases, liquids, and solids. The amount of heat that is present will affect what condition matter is in. Take water for example. When it's frozen, there is an absence of heat; as it gets warmer, it melts and turns into a liquid; warmer still, and it turns into steam.

What is important for our study of cosmology is that out of the study of heat would come the laws of *thermodynamics,* and these laws play a very important role in understanding the origin, evolution, and eventual end of the universe. I'll go into these laws for you later when we discuss present-day cosmology.

Universal Constants

Energy in science is simply the ability to do work. And work is the transfer of force from one body or system to another.

Universal Constants

Thermodynamics involves the study of the reversible transformation of heat into other forms of energy, such as mechanical energy, and also covers the laws governing those transformations.

The Dynamics of Electromagnetism

Ancient people knew of stones that attracted certain metals called magnetite, a form of iron ore. They also had some unusual remedies that involved magnetite. Supposedly it was good for curing arthritis, gout, cramps, and baldness. (I wonder if the Hair Club knows this?) But no serious scientific work involving magnetism occurred until Henry Gilbert theorized that the reason a compass needle points north and south is because the earth has a magnetic field.

The study of magnetism is a significant branch of science today. Breakthroughs in research have benefited medicine, industry, and particle physics. But 200 years ago most of what came to be known about it was due to the investigation of the properties of electricity.

Universal Constants

If you've seen the movie *Jurassic Park*, you know what **amber** is. That was the stuff that the mosquito that carried the dinosaur DNA was trapped in. In reality it's a form of petrified, fossilized tree resin.

Static electricity was the first form of electricity discovered. Some people thought it was a form of magnetism because when you rub *amber* or a glass rod with silk or wool it will attract little pieces of paper or other light objects. It's what causes static cling on your clothes and hair. Experiments with static electricity revealed that it was a different form of electricity. It was static, or in other words it didn't move. The electricity we're most familiar with is based on the movement of electrons through a medium, such as a wire, from a negative to a positive pole.

There were a number of individuals who contributed to the understanding of electricity and its relationship to magnetism. The following list is some of the most important scientists and what they discovered:

➤ Alessandro Volta built the first battery, called a voltaic pile. He discovered something called EMF, or electromotive force. You know at as volts. It was named after …?

➤ Hans Christian Orstead did the first experiments with magnetism and electricity. He discovered that when you put a compass next to a coil of wire that's connected to a power source, such as a battery, the compass points toward the wire rather than to magnetic north. He coined the term electromagnetism.

➤ André-Marie Ampére did the math and came up with the law named after him. It defined the relationship between a magnetic field and the amount of current. He explained that electric currents produce magnetic fields. Oh, and the unit for current is, need I say … ampere, or amps.

➤ George Simon Ohm experimented with various substances of different thickness to see which ones conducted electricity better than others. He discovered that some materials carried electricity better than others, while others resisted its flow and hence the term resistance was coined. The unit for resistance is called an ohm, and *Ohm's Law* was born.

Universal Constants

Ohm's Law defines the relationship between voltage, current, and resistance. If you know this law you understand most of what electronic theory is based on. Here's how it breaks down:

➤ Voltage is equal to current times resistance, or $V = I \times R$

➤ Current is equal to voltage divided by resistance, or $I = V \div R$

➤ Resistance is equal to voltage divided by current, or $R = V \div I$

➤ Michael Faraday discovered inductance. He already knew that electricity could produce magnetism, and he wondered if the reverse was true as well. His experiments showed that when he passed a magnet through a coil of wire, the movement produced a current that registered on the meter he had hooked up to it. This was called inductance, because the moving magnet induced a current in the coil of wire. Faraday also explained, for the first time, how this process worked by developing what is called a field theory. If you've ever sprinkled iron filings on a piece of paper and placed a magnet under it, you can see the magnetic lines of force that comprise the field. Faraday used this concept to explain how electricity was induced in a coil.

With the knowledge gained from all of these experiments, the study of electromagnetism soon revealed some astounding things about the structure and nature of the universe. That's coming up in Part 3, "And in This Corner ...."

Light, Lenses, and Optics

The fourth category that was crucial to the development of twentieth-century cosmology was the study of optics. Part of optics has to do with the study of the human eye and how it sees. That sounds like a strange area of study to include as one of our categories, but it's what was learned about how vision operates and its relationship to light that was important.

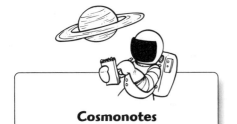

Cosmonotes

The study of optics deals with the nature and properties of vision and light. It has led to developments in research such as the use of lasers for eye surgery, new lenses for glasses and contacts, and fiber optics.

Inquiry into human vision is as old as (who else?) Aristotle. The structure of the eye has always been intimately linked to the nature of light. For without light, there is nothing to see. The ancient philosophers thought that people were able to see because the eye was the source of its own light. This idea changed over time as medical knowledge revealed the true structure of the eye.

The breakthrough came when the phenomenon of *refraction* was finally understood. This understanding coupled with the medical knowledge gained about the structure of the eye led to the development of lenses. And from that we got eyeglasses, the telescope, and the microscope. Along the way, since light played such an important part of all of this, questions were raised about the nature of light.

Universal Constants

Have you ever put a pencil in a glass of water and noticed that it looked like it was broken? The piece sticking out of the water and the piece in the water don't continue in a straight line. That's due to the phenomenon called **refraction.** This occurs when light passes through mediums of different density, such as air and water, and these bend the rays of light in different directions. It is what allows the focusing of light through curved surfaces, such as a lens.

Isaac Newton made significant contributions to the study of optics, besides all of his other accomplishments. The following is a brief summary of some of his discoveries and theories:

➤ Sunlight, or white light, is composed of all of the colors of the rainbow. You can use a prism to separate out the colors.

➤ Light is made up of particles that move through space in straight lines.

➤ When the particles of light come in contact with a surface, they cause the particles of the material to vibrate and that produces the distinct and various colors we see.

➤ These particles all move at a constant speed.

Another scientist, Christian Huygens, disagreed with Newton's particle theory. He believed that light was a wave and gave some good hard evidence to support this. But because of Newton's status, not too many people accepted this idea. Later on you'll find out that the two were both right.

Just a Little Bit of Chemistry

As you already know, one of the most direct results of the alchemical tradition was the new science of chemistry. What was learned from years of research in chemistry would be a key component in the development of twentieth-century cosmology, especially as it relates to the microcosm. As before, here is an overview of some of the most significant discoveries made by a handful of chemists:

➤ Remember our buddy Robert Boyle? Well, he proposed the idea that an element was the simplest form in which a substance could exist.

➤ Robert Black, Joseph Priestly, and Daniel Rutherford through their study of gases all discovered the separate composition of air: Black, carbon dioxide; Priestly, oxygen; and Rutherford, nitrogen.

➤ Antoine Lavoisier developed the idea that matter was composed of different kinds of elements.

➤ John Dalton realized that these elements combined in different ways to produce compounds, such as water, salt, alcohol, etc. From this he came up with the idea that elements are really composed of smaller particles called atoms, and each atom has its own unique atomic weight.

Universal Constants

The **periodic table of elements** provides an elegant and systematized way of ordering all of the known elements in the universe. It's one of the great achievements in science. Today there are 113 known elements, many of which are created in nuclear reactions where they exist for only an instant before they disappear.

➤ And finally we have Dmitri Mendeleev. He composed all of the known elements at the time, which were 69, into the *periodic table of elements*. This table categorized each element by similar properties and atomic weight.

The discoveries in chemistry led to further investigations into the building blocks of the microcosm. The knowledge acquired was combined with electromagnetic research, the nature of light, and Newtonian cosmology to provide a comprehensive explanation for the structure of the universe, or so it was thought. Before we leave this chapter, let's look at one more contribution by Newton—the universal law of gravitation.

The Attraction of Gravity

Needless to say, there are few individuals who made as many discoveries or contributed as much to science as Newton did. Besides everything else I've discussed, he also did a few more amazing things. His three laws of motion accounted for the interaction of all earthly physical objects by clearly defining their movement in new concepts and terms. Words such as mass, acceleration, force, velocity, and momentum became common vocabulary to explain his new laws.

One of the most important laws is the *universal law of gravitation*. Newton's calculus allowed him to describe in mathematical terms Kepler's laws of planetary motion. He was able to calculate the masses of the Sun, Moon, and planets based on the mass of the Earth. He explained the exact motion of the moon and the tides of the oceans. His gravitational law explained the entire workings of the solar system.

Universal Constants

The **universal law of gravitation,** without actually getting into the formula, simply states that objects with higher masses have a greater force of attraction, as do objects that are closer together. And for those of you more mathematically inclined it's explained: The force between two objects is proportional to the product of their masses and inversely proportional to the square of the distance between them. Got it?

I'll come back to Newton's cosmology a few more times over the next couple of chapters, no sense it giving it all away at once. We will be getting into the larger domain of cosmology, the macrocosm, fairly soon and I'll want to show you how some of the key concepts of Newtonian cosmology would have to be changed. Coming up: the dual nature of light, Einstein's theories of relativity, and the quantum universe. Oh yes, and we'll see how God fits into the picture, too.

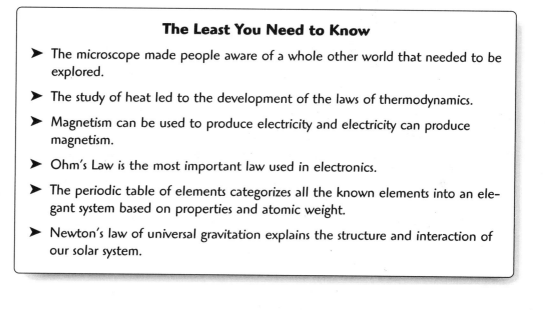

The Least You Need to Know

➤ The microscope made people aware of a whole other world that needed to be explored.

➤ The study of heat led to the development of the laws of thermodynamics.

➤ Magnetism can be used to produce electricity and electricity can produce magnetism.

➤ Ohm's Law is the most important law used in electronics.

➤ The periodic table of elements categorizes all the known elements into an elegant system based on properties and atomic weight.

➤ Newton's law of universal gravitation explains the structure and interaction of our solar system.

Part 3

And in This Corner ...

Newton's three laws of motion along with the law of universal gravitation explained in very precise mechanical ways the clock-like structure of the universe. Contained within this cosmology was the belief that space and time were absolutes. In other words, they were constant, never changing dimensions against which other types of physical phenomena could be measured. They were considered universal constants.

This view continued into the early part of the twentieth century, until Albert Einstein came along and showed that both of these universal constants were relative and not absolute dimensions. Besides Einstein altering Newtonian cosmology, there were other changes in how the universe was structured, especially our place in it. Darwin's theory of evolution questioned the religious views of creation, and physicists came up with a couple of their own theories on the origin of the universe. Let's take a closer look at some of these theories to see how relative everything can be.

Chapter 9

Evolution vs. Creationism

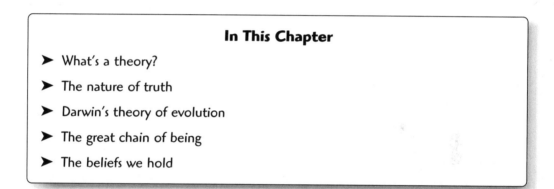

In This Chapter

➤ What's a theory?

➤ The nature of truth

➤ Darwin's theory of evolution

➤ The great chain of being

➤ The beliefs we hold

In this chapter, I will be discussing two aspects of cosmology that have been, at times, antagonistic to each other. One is the theory of evolution and the other is the belief in creationism. While a summary of each will be provided, some important questions regarding both systems of thought will be included. Anytime topics such as this are discussed it's important to keep an open mind and approach the material nonjudgmentally.

As you know, this book is about theories of the universe. Let's pause and do some reflective thinking for a moment. What exactly is a theory? We use the word in various ways, ranging from simply contemplating or speculating about certain things, to formulations about apparent relationships and underlying principles of observed phenomena. This last definition is normally how science uses the word, but in this sense, there are theories that can be equally as good as scientific ones.

One of the things that makes a theory acceptable at all is the amount of evidence there is to support the theory. But evidence can come in many ways other than strictly scientific. Just because a theory is scientific that doesn't make it truer than theories that aren't. A theory is simply that, a theory. It's not an absolute, nor is it proof that what is being discussed is the way things really are. It is open to questioning, refutation, and disproof. It can also lead to better theories, deeper insights, and a closer approximation of what is really true. And that is the crux of the whole thing, isn't it. What is the truth?

Is Truth in the Eye of the Beholder?

In the introduction to this book I made the statement, "Just because you believe something is true doesn't mean it really is." I don't plan on taking you through three thousand years of philosophical debate about what the nature of truth is. (You can find that in *The Complete Idiot's Guide to the Truth,* just kidding.) But in order to clarify my statement, let's look at some definitions of the word true.

➤ Something that is true is considered to be any and all of the following: correct, right, certain, accurate, reliable, actual, real, and constant.

➤ What is true also agrees with fact, experience, or reality.

➤ And the truth is just an extension of what is true.

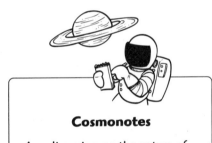

Cosmonotes

Any discussion on the nature of truth has to involve how we gain knowledge of it in the first place. This search for how we know anything at all falls into a realm of philosophy called epistemology. If this is of interest to you, any good introductory text to philosophy can get you started in the area.

So truth, or what is true, based on the definitions above, can be divided into two types: subjective and objective truth. And this is where we run into problems. Some people argue that there is no such thing as objective truth. Since everything outside of us is merely filtered through our perceptions and beliefs, we see what we want to see.

Then, of course, there is the other position that argues that all we can really know are objective truths (truths apart from our perceptions of the world). They exist whether or not we perceive them. Into this category would fall facts such as $2 + 2 = 4$, a triangle has three sides, and other self-evident mathematical concepts. The scientifically minded individual often takes the position that there are objective truths because for them what science has shown to be true and ultimately real can't be disputed because of the powerful methodology found in science.

But whether it's the scientific method or the filtering of reality through our perceptions and beliefs, it all comes back to us. Truth is only meaningful to us. It's a human concept. If the truth is out there (*X-Files* music in the background) or inside of us, to what degree does that matter? We are the ones that accept or reject what others think, feel, or believe. It doesn't matter if it's religion, science, politics, or cosmology, we live our lives based on our own unique version of what is true for us.

Mindwarps

The laws of physics are regarded as truths about the natural world. If a law is broken it doesn't mean that it's no longer true, it just means that more information is required so that it can be revised to include the event that broke it in the first place. It then becomes a more encompassing law containing a higher element of truth. This same process can be applied to what we believe to be true as well.

It's a Matter of Opinion

At this point I want to include two of my favorite quotes from Albert Einstein. They're very appropriate for this section on truth:

> Every man has his own cosmology and who can say that his own is right.

> Whoever undertakes to set himself up as the judge in the field of Truth and Knowledge is shipwrecked by the laughter of the gods.

So is that it? Is truth only in the eye of the beholder? What is true for me may not be true for you? Truth is just a relative thing? If that's the case, is it really the "truth"? Let's take a few examples to see if we can find out.

If you ask a Moslem or a Christian whose religion is the true religion and they both reply that theirs is, then which one is true? Can they both be right? If each claims that what they believe is the truth, how can it be the "truth"? How do you resolve this type of dilemma? I'll come back to that question in a little while. Let's look at another example.

What about politics? Of course many of us might reply that there is no truth in politics, so why even bother addressing this? Okay, then let's talk about political opinions, which may be as close as we can get to discussing truth in politics. The central force motivating a democratic society is supposed to be the "will of the people." Our elected representatives are considered to be the voice of the people. Is this the truth? Have they been our voice? Some people think they are, others think they aren't. Everyone has an opinion. But are these opinions based on true knowledge? Are they *informed* opinions? In many cases they aren't. As humans we have a tendency to react to things from an emotional place, and emotions can sometimes have very little to do with what is true. Emotions can and do obstruct clear, focused, critical thinking and will potentially prevent an understanding of what the truth really is. (And this has nothing to do with men being from Mars and women from Venus.)

Universal Constants

Thomas Jefferson once said, "The only good political opinion is an **informed** political opinion." To be informed is to have genuine knowledge of the issue at hand based on re-search, facts, and firsthand knowledge. In other words, you have a good handle on the truth. Like Socrates said, "The Wise person is the one who knows that they don't know, while the fool will claim knowledge about that which they really know nothing."

That's not to say that there is anything wrong with emotions. I'm not advocating being Mr. Spock from *Star Trek.* (I find that a rather amusing statement, Captain.) But what I am saying is that emotions have their place and function, but they don't al-ways work very well in trying to ascertain the truth when it comes to issues that we are too emotionally close to.

Degrees of Truth

To get back to the original question of truth being in the eye of the beholder, instead of truth being relative, many think that there are varying degrees of truth. What do you think? Perhaps there is a point where subjective and objective truths meet? If that is so, then there must be an intimate connection between truth, belief, under-standing, and our experience of reality. Maybe the division between subject and object—or in other words, between ourselves and the world around us—is really not a division at all. Experts in the field of human consciousness research suggest that the experience of subject and object is only a reflection of a specific state of conscious-ness and that we have the potential to experience the world in other ways than how we usually do. This suggestion and the accompanying theories will be explored in more detail in chapters to come. For now, let's take a closer look at degrees of truth.

Why do we consider something to be true? The only source we have for that is our experience in the world, our daily lives. Truth is ultimately based on our beliefs, and our beliefs are structured from our life experiences. This seems simple enough, but it also has numerous implications. Since each of us has had our own unique life experiences, the beliefs we have all developed are also, in the end, just our own. It's the level and application of understanding that has or hasn't become part of our paradigm that reveals truths to be of a lesser or greater degree. Understanding, like truth, seems to manifest in degrees of comprehension. Which statement do you think reflects a greater measure of understanding: "My religion is the only true religion" or "all religions are true"?

If you believe that scientific knowledge is the only true source of knowledge about the world, that doesn't leave much room to acquire knowledge from other areas. The realization of the nature of truth as a dynamic, expanding process goes hand in hand with increased understanding. And both just don't happen, each has to be sought after. If truth becomes fixed and crystallized, it will break like a dead branch when the winds of change blow.

Cosmonotes

The experience of truth was equated by Greek philosophers to the experience of beauty. To know what is true, one needs to experience it. This is not an emotional experience, nor is it strictly a mental one either. It's actually a rather difficult occurrence to explain. It's been described as thinking with your heart or feeling with your mind. It involves a level of knowing that transcends either the emotional or the mental by themselves and is more a combination of both. For lack of a better term, it's also been called inspired intuition or direct cognition. How often has that occurred in your life?

How Did We Get Here?

I wanted to spend some time discussing ideas about the nature of truth because so much of what we believe about the world, the universe, cosmology, and God is based on what we consider to be true. At the end of the nineteenth century, many scientists believed that all that could be known about cosmology was indeed known, and any anomalies would soon be resolved. Little did they know what Einstein and the rest of twentieth-century physics had in store for them.

Black Holes

A common misconception in evolution theory is the basis for the process known as natural selection. Many philosophers and scientists have interpreted it to mean that the survival of a species is based on those most fit, or it's "the strong that survive." In reality this is far from the truth. Survival is based on adaptability to change and has nothing to do with the strong overcoming the weak. As your environment changes, it all has to do with how well you can adapt to the changing conditions. And this idea can be applied to many areas of our lives, too.

What was to happen to nineteenth-century Newtonian cosmology had already occurred to some degree with the accepted story of creation. While science continued in the twentieth century to expand upon its theories, religious dogmatism fought any change to its interpretation of the truth. What was it that threatened the long-held belief of creation?

You Have Been Selected

Charles Darwin (1809–1882) published his *On the Origin of Species* in 1859 and set forth his theory that animals evolved through variation and natural selection to those most fit to survive in particular environments. Here's a brief summary of some of its main ideas.

➤ Biological organisms and species do not have a fixed, static existence but exist in permanent states of flux and change.

➤ All life, from a biological point of view, takes the form of a struggle to exist, and to exist to produce the greatest number of offspring

➤ This struggle for existence removes those organisms less well adapted to any particular ecological system and allows those better adapted to flourish. This is the process called natural selection.

➤ Natural selection, development, and evolution require enormously long periods of time—so long that the everyday experience of human beings provides them with no ability to interpret these histories.

➤ The genetic variations that produce increased survivability are random and not caused by God or by the organism's own striving for perfection.

Cosmonotes

While Darwin's theory is far from being complete, and some scientists would be the first to admit that, it does stimulate thinking and breeds new theories. That is its strength. Its weakness, besides being incomplete, is that it, like most of science, supplies no purposeful action for the underlying phenomena it seeks to explain. Isn't that as important of a question to ask as the questioning of religious dogma?

The Ladder of Creation

The effect of these points was to do the same thing that the Copernican revolution did, and that was to move humanity away from the center of creation and imply that it could hardly be its crowning glory. Natural selection, which apparently leaves no place for God in the world, has proved the most difficult part of the theory for some to accept. What other Biblical traditions does it question? Well here's a few that pop out:

➤ By emphasizing that species changed, evolutionary theories apparently destroyed ancient notions of the *Great Chain of Being,* in which all living organisms had their proper place in a fixed, immutable order.

➤ By emphasizing that species changed over time, the theory called into question the literal truth of the Bible, which gave a much shorter span of time from the creation to the present.

➤ The displacement of humanity from the center of creation along with our recent arrival in geological time brought our importance to a questionable level.

➤ Natural selection, with its notions of randomness and apparently wasteful cruelty of the selection process, argued against any form of divine morality.

If this wasn't enough to get the pot boiling, in 1871 Darwin published *Descent of Man.* This work more than any other brought into question humanity's creation because the main thesis of the theory was that in our long development we had originally evolved from primates (you know, gorillas, chimpanzees, apes). Here's a brief excerpt from it. Original sources can't be beat.

The main conclusion here arrived at, and now held by many naturalists who are well competent to form a sound judgment, is that man is descended from some less highly organized form. The grounds upon which this conclusion rests will never be shaken, for the close similarity between man and the lower animals in embryonic development, as well as in innumerable points of structure and constitution, both high and of trifling importance, the rudiments which he retains, and the abnormal revisions to which he is occasionally liable, are facts which cannot be disputed. They have long been known, but until recently they told us nothing with respect to the origin of man. Now when viewed in the light of the whole organic world their meaning is unmistakable. The great principle of evolution stands up clear and firm, when these groups of facts are considered in connection with others, such as mutual affinities of the members of the same group, their geographical distribution in past and present times, and their geological succession. It is incredible that all of these facts should speak falsely. He who is not content to look, like a savage, at the phenomena of nature as disconnected, cannot any longer believe that man is a work of separate act of creation.

Universal Constants

The **Great Chain of Being** or the Ladder of Creation has been a long-standing belief that explains the natural order of the world in a hierarchical scale. It's based on the Book of Genesis account of creation in which God created all of the various species of plants and animals fully formed and distinct. It's a belief system that also developed out of the Aristotelian classifications of the flora and fauna of earth. You could even call it a creation theory that links all forms of life from the lowest to the highest, and, of course, the highest being man. Hmm ... any sense of self-importance there?

There you have it, the grounds for the great debate that still rages in our society today. If you didn't before, I think you now have a general idea of early evolution theory. Let's look at the other side of the coin and see what creationism has to say.

And God Said ...

We've already discussed some of the creation story found in Genesis as well as its source. But that is just part of the whole belief system known as *creationism.* Besides the creation of the universe and earth, there is also the creation of man, specifically Adam, in God's image and the entire story that takes place with Eve in the Garden of Eden.

Universal Constants

Creationism is a term that applies to individuals and groups who believe in the literal word of the Bible and the entire creation story. However, there are many different groups of creationists. There is scientific creationism, which seeks to use science as a means to disprove the theory of evolution. There are also forms of liberal creationism that accept some of the Bible and some of evolution theory as well. Then there is pure creationism that accepts only the Biblical portrayal and believes that the Bible is the literal word of God.

Creationism has a wide variety of adherents, from scientists to fundamentalists. Regardless of the many interpretations it has, creationism does have as its core the essential belief that God created the world and everything in it. After that the various arguments and proofs cover a wide range of beliefs and theories.

Emotions in support of creationism often run high because the belief in God is so much a part of our society. But a belief in God doesn't mean that what is in the Bible is the only true source regarding creation. Only a few years ago, the Pope came out in support of science's views on evolution. He believed that there was no reason why evolution couldn't be part of God's overall plan. For him and many of his followers, God provides the purpose and the reason for life unfolding as it does.

But as with anything there will always be people who claim that what they believe to be true, should be everyone's truth as well. So it all comes back to the question of truth once more, doesn't it? It also seems to be part of human nature to tenaciously pursue and impress upon others the uniqueness of one truth over another.

Mindwarps

Have you ever heard of the Scopes Monkey trial? On July 25th, 1925, a biology teacher by the name of John Scopes was put on trial in Tennessee for teaching the theory of evolution. Clarence Darrow, a brilliant lawyer, was leader of the defense and William Jennings Bryant—three-time presidential candidate, Secretary of State to Woodrow Wilson, and evangelist—was the leader for the prosecution. It's considered one of the greatest courtroom examinations ever, and revealed the conflict that exists between creationism and evolution. The trial was made into a Broadway play and the movie, *Inherit the Wind*.

To Believe or Not To Believe, That Is the Question

A point that needs to be made before we go any farther, is that many of the questions raised and points made in this last section on beliefs comes from discussions among philosophers, scientists, sociologists, theologians, and historians of science. It's not meant to be preachy or to come across as discrediting anyone's particular beliefs. Only that there is always more to the development of understanding than we sometimes assume.

Mindwarps

One of many interesting recent theories in the area of human consciousness research has shown that we are probably hardwired in our brains to look for what is meaningful to us in life. It is what seems to be at the core of our innate need to believe in something upon which we can base our lives. So does it mean that some of us may be more inclined toward a religious rather than scientific point of view and vice versa? Possibly, and that in and of itself raises some interesting questions. Maybe some of us weren't designed to believe in God and others were. (Would God do that to us?)

Why do you think that we have this incessant need to convince others of the correctness of our point of view? Many individuals think the problem lies in the degree to which we identify with our beliefs. Under the guise of rationality or authority or both, beliefs are used as a means to argue for superiority of correctness. But we can't see the forest for the trees, because we can become blinded by our inability to see the world other than the way we want to see it. We become so strongly identified with what we believe to be the truth, that it's impossible for us to even consider that we could be wrong, or at least misinformed.

Beliefs are a lot like theories, they can go a long way in explaining why something is the way it is, but it's not an absolute. A belief, again like a theory, is just that—a belief. Are we only what our beliefs are? Perhaps not. When you come to think of it, we're really not aware of any one particular belief most of the time in our daily activities. It only comes into play when something evokes a response from it. And then, of course, we're ready to take whatever action is necessary. Now we're so identified with it that an attack against or a support of the belief gets an immediate response in kind. (This may be a little abstract, so just remember the next time someone questions your belief about something; see how you react.)

Have you ever changed your belief about something sometime in your life? If so, what happened? How did you feel or think differently? Are you still the same person you were before, or are you different because your belief changed? No answer is required here; it's just something for you to think about.

Under all of this lies the fundamental question, why do you believe what you do? If you accept the authority of science or religion, how do you really know that what they are telling you is really true? Have you experienced that truth for yourself? Of course, you can't go around questioning every single element of either of these systems, but at the same time should you swallow whole the admonitions of religion and the purely materialistic claims of science? The ability to *discriminate* can go a long way in helping us to arrive at our own conclusions about things.

Over the years science has sought to become more popular in the eyes of the general public. That's a great idea because it helps to demystify some aspects of science and makes many of the ideas more accessible to you and me. But at the same time, science hype has also increased. Headlines such as "Science Now Knows How the Universe Will End," or "The Smallest Particle That Can Exist Has Finally Been Found," come across as fact rather than theory. It's a misleading way to attract people to buy books and magazines. Hyped up advertisement also only goes to reinforce the assumption that only science can solve the questions that we are asking. If science has learned anything, it's that

Universal Constants

To **discriminate** means to distinguish between things, to recognize a difference. And a discriminating mind is one of the most powerful tools you can have to arrive at your own understanding of things.

just when you're sure you have the final answer, a new theory, idea, or solution makes itself known. Instead of seeing the discoveries of science as stepping stones to greater understanding, popularizers of science and even scientists often consider the goals of science as ends in themselves.

The main ingredient missing in all of the answers that science and religion have to offer is simply asking why. Not how, or when, or who, but why. And that really puts the whole thing back on us. Human consciousness—what it is, why it is, and how it operates—plays a crucial role in our understanding of anything, and has to be included in any formulation, theory, or explanation that either science or religion has to offer. Our understanding of what consciousness is is still in the early stages. What is needed is a Copernican revolution in the field of human consciousness that will help to explain consciousness's role in the cosmology of the universe. It may even help explain why.

In upcoming chapters we'll explore the role of human consciousness more in depth, especially in relation to the quantum world. But we have a few more cosmological seeds to plant before we get there. Next stop, some important discoveries about light and other relatively new ideas.

The Least You Need to Know

➤ What is true is based on fact, your experience, and your beliefs.

➤ Many people think there are two kinds of truth, subjective truth and objective truth, and the nature of truth had been discussed for centuries.

➤ The theory of evolution is based on the writings of Charles Darwin and has become the most widely accepted theory in science for the origin of humanity.

➤ Creationism is a system of religious belief that considers the Bible to be the truest explanation for the creation of humanity and the universe.

➤ It's important to consider the role of our consciousness when we come up with scientific or religious answers to questions that we have.

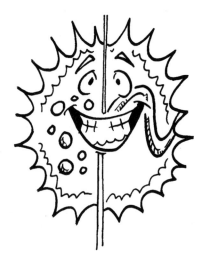

The Dual Nature of Light

Light enables us to see and comprehend the universe and is an important part of any study of cosmology. But what kind of thing is it? While it is first and foremost a sensation in the eye, it also has an independent existence outside of us. Newton thought that it was a beam of particles, but what kind of particles? What are they made of? What size are they? What shape? These questions went unanswered until Einstein and a few other physicists arrived on the scene.

By the end of the nineteenth century everyone had conceded that Newton was wrong and that light was a wave, but what kind of wave? An ocean wave is not a thing, it is a property of water, something that water does. If there is no water there is no wave. So if light was a wave, what was waving? This was the most urgent question that physicists were asking. By the time an adequate answer was found, light would end up being described as both a particle and a wave. Yet how could it be both? This would be the first of many paradoxes that would begin to question our commonsense notion of how the universe operates.

Waves of Light

While the nineteenth century was filled with many noteworthy scientific discoveries, one of the most significant was the description of light and its properties. The insights that the study of light provided were later crucial to quite a few of Einstein's theories and also formed the basis upon which the big bang theory developed. The key realization that led to all of this was that light was an electromagnetic wave. Let's take a look at how this discovery was made.

About 100 years after Newton's theory about the particle structure of light, a man by the name of Thomas Young performed a very famous experiment in which he showed that light *propagated* as a wave.

In Young's experiment, a light source was shown on a screen that had two holes a few millimeters apart. He put another screen behind the first, and the light coming through the two holes of the first screen illuminated this second target screen. As expected, two patches of light appeared. He then made the holes smaller and the corresponding patch of light became smaller, too.

But then something very unusual happened. When Young made the holes very small, faint rings appeared around the patches on the target screen that actually made them bigger. Instead of the patches reducing in size to correspond with the smaller holes, they were larger. This couldn't happen if light were made of particles, because particles move in straight lines and wouldn't make these larger faint rings of light around the patches.

If he made the holes even smaller, the patches of light on the target screen began to overlap and became crossed with dark lines. These dark lines were caused by waves of light interfering with one another.

Universal Constants

Propagate is a term used in both biology and physics. In biology, it refers to the reproduction of a species while in physics it refers to the transmission of sound waves or electromagnetic waves through air or water.

You can get a good idea of just how this works if you drop a couple of rocks in water. The ripples sent out by each rock hitting the water interact with each other. Some cancel each other out and some amplify each other. The light and dark bands seen on the screen are the result of light waves doing the same thing. The dark band is the absence of light, or when light waves cancel each other out, while the lighter bands are where light waves amplify each other. When light waves are forced to travel through very small areas, like the two tiny holes in Young's experiment, the interference pattern that is created can only result if light is a wave, not a particle. Although many scientists scoffed at Young's experiment he was later exonerated by the work of two other physicists, James Maxwell and Heinrich Hertz.

You Can't See It, but It Has to Be There

An important question to ask about light behaving as a wave is what does it wave in? The idea that there was some type of invisible fluid or element that was part of the universe can be dated all the way back to Aristotle. Remember the four Greek elements: earth, air, fire, and water? It was thought that there was a fifth one that helped to explain the movement of the celestial bodies in the heavens, because it was believed that something was needed to support and guide them in their perfect circles.

Mindwarps

The belief in the existence of a substance called ether (not the anesthetic) dates back to the ancient Greeks. It was originally believed to be some sort of material that permeated all of creation. Many scientists believed it existed up until the end of the nineteenth century. Today it is often associated with the energy or life force found in traditional Chinese medicine and the Ayurvedic medical practices of India. In China it's called Chi and in India it's called Prana. In esoteric philosophy it is thought to make up the etheric body or energy body that is the life force that animates our physical bodies.

In a speech delivered in 1889, Heinrich Hertz said:

> … the great problem of the nature and properties of the ether which fills space, of its structure, of its rest or motion, of its finite or infinite extent. More and more we feel that this is the all-important problem, and that its solution will not only reveal to us the nature of what used to be called imponderables, but also the nature of matter itself and of its most essential properties—weight and inertia …. These are the ultimate problems of physical science, the icy summits of the loftiest range.

That statement really reflected how many prominent scientists felt. Isaac Newton along with Michael Faraday and James Clerk Maxwell all believed that there had to be some substance that acted as a medium to transmit the forces of gravity, electricity, and electromagnetic waves through space. However, they had different views on exactly what this substance was. Newton himself had no idea what it was, as seen in this letter to a friend.

... that gravity should be innate, inherent and essential to matter, so that one body may act upon another at a distance through a vacuum, without the mediation of anything else, by and through which their action and force may be conveyed from one to another, is to me so great an absurdity that I believe no man who has in philosophical matters a competent faculty of thinking can ever fall into it. Gravity must be caused by an agent acting constantly according to certain laws, but whether this agent be material or immaterial, I have left to the consideration of my readers.

Faraday's Fields

Remember Michael Faraday from Chapter 8, "Physics 101"? As an experimenter Faraday was extraordinary. His ideas were unlike anybody else's and he had little to do with the experts at the universities. He was an intuitive thinker and many of the ideas that he expressed were difficult for the typically rational approach of many of the other scientists. He was not that good at the higher mathematics of algebra and calculus, so he worked out his theories strictly through experimentation rather than mathematical equations.

Cosmonotes

Michael Faraday only had a formal education through the eighth grade and had no interest in mathematics. His brilliance laid in his innate ability to put his ideas into experiments. Over the years, he invented the first electric motor, showed that magnetism could be converted into electricity, designed and built the first dynamo, and made the first electrical transformer. He became the Grand Old Man of English science, but his contemporaries read his papers more for the results he found rather then what they meant. He was a strong believer in the unity of all types of physical interaction, a belief that was also fundamental to Einstein and which lies at the core of the search for a unified theory today.

According to Faraday, matter is atomic, which was something that most scientists agreed upon. Each atom had an "atmosphere of force" around it, an idea that had never been clearly defined and had no better name. He later developed this idea into the term "field," a form of ether that had lines of force running through it. As mentioned before, the best way to see this is to put iron filings on a piece of paper and place a magnet under the paper. You will see a pattern in the filings created by the magnetic field. Here again was the idea that this field had to exist in something; it couldn't just float about in empty space.

After years of experimentation, Faraday's discoveries suggested that electric and magnetic fields were more than just forces; they also had a dynamic interaction. This capacity to use one to create the other led to the term "electromagnetic field." And perhaps his greatest insight was that an electromagnetic field could affect a ray of *polarized* light. He realized that there had to be some connection between electromagnetism and light waves.

Universal Constants

Polarized light is a form of light in which the electromagnetic waves of light move only in one plane. Normally they move at right angles to each other. If you don't quite understand this, think of a normal light wave as a rotating barbershop pole. The spiraling coloring lines are moving around and around in three dimensions. Now if you flatten the pole out, you just have lines moving zig-zag up and down on a flat piece of paper. A polaroid filter does the same thing. It only allows waves of light to move up and down, or left and right, rather than around and around. (This has nothing to do with the Polaroid camera; they just liked the name.)

Maxwell's Mathematics

The man who would bring electromagnetism and light together was James Clerk Maxwell. He developed mathematical formulas that connect electric fields with electric charges, magnetic fields with electric currents, and those that connect electric and magnetic fields together. These are the formulas that express the mathematical unification of electricity and magnetism. There is not a word about the properties of ether in all of this; he was only concerned with fields and their sources. However, he was still an ardent believer in it as can be seen in this article he wrote for the *Encyclopedia Britannica:*

> Whatever difficulties we may have in forming a consistent idea of the constitution of ether, there can be no doubt that the interplanetary and interstellar spaces are not empty, but are occupied by a material substance or body, which is certainly the largest and probably the most uniform body of which we have any knowledge.

Cosmonotes

The existence of ether is no longer one of the hot topics in science today. It was disproved by some experiments done in the nineteenth century, but these were set up to measure the existence of ether in just a certain way. Albert Einstein didn't believe that there was any problem with his theory on general relativity and the co-existence of ether. Whether it really exists or not has never really been proven. When I discuss general relativity in a few chapters, we'll come back and look at ether again to see how it may relate to properties of space.

Maxwell, like Newton and Faraday, believed that the existence of ether was necessary, but like them, he had no clue as to what it really was. He kept his views about ether out of his mathematical formulas, not just because he didn't know what it was, but because it wasn't necessary in developing his mathematical formulas. However, his belief in ether did lie behind his theories of how electromagnetic waves moved through space.

Maxwell's theories explain so many different facts of electricity and magnetism that there is no serious doubt that they are accurate. His mathematical formulas are the cornerstone of classical physics in electromagnetic radiation. From the work that he did, he was able to calculate the speed of electromagnetic waves, and because the speed of light (which had been discovered by Léon Foucoult) was essentially the same, he deduced that light was electromagnetic, although he gave no experimental evidence to support this.

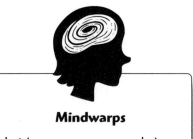

Mindwarps

Just in case you were wondering or didn't know what the speed of light and all electromagnetic waves was, it is 186,000 miles per second, or 299,800 kilometers per second, or 670 million miles per hour. This speed is constant; it never changes and is the universal measuring tool to establish the distance of all celestial bodies from earth.

The Experiment That Failed

Now that the speed of light had been accurately measured, the question of the existence of ether still needed to be addressed. In 1887, a famous experiment was set up to determine if the presence of ether could be found. It was called the Michelson-Morley experiment, named after the two scientists who set it up. The reasoning behind the experiment went something like this. If ether is motionless in space, then any object moving through it, like the Earth, would encounter an ether wind blowing in the opposite direction. The Earth moves through space as it orbits the Sun. Physicists thought that the ether wind was blowing past the Earth at the same speed that the Earth orbited the sun.

The speed of this ether wind could be measured by using a light beam, since the speed of light is always constant. (I'll cover how they knew that in Chapter 11, "Cracks in a Newtonian World.") The idea was to measure the motion of the Earth with respect to the fixed ether by measuring the speed of light as it moved back and forth in different directions on the surface of the Earth. The ether wind would affect the speed of light waves, just like wind affects the speed of sound waves in the air. According to this theory, the speed of light would increase in some directions and decrease in other directions. When the various speeds were compared, it would be easy to calculate the speed of the Earth. Since the speed of light is a fixed velocity, any fluctuations in its speed would be accounted for by the ether wind.

The experiment was set up very precisely, and carried out a number of times, but it was an utter failure. There was no ether wind discovered. But this failure ended up changing the face of physics. It forced physicists to rethink their ideas about the structure of the universe and this long-held cosmological belief was soon regarded as a quaint superstition. However, even though this particular experiment didn't show any sign of ether, it didn't stop some scientists from still believing in its existence. And an important point to remember is that any experiment is designed so that certain outcomes, within particular parameters, may or may not be found. Just because this particular experiment didn't show any signs of its existence, doesn't mean that another theory or experiment couldn't. The existence of ether still has yet to be solidly proved or disproved. It's just fallen to the side, because its existence isn't necessary for the many theories and mathematics of today's cosmology.

Universal Constants

Radiation is a process that occurs when the atoms and molecules within any form of energy undergo an internal change and is sent out as rays or waves through space. For example, you can feel heat radiating from a stove when you stand near it. Or when you stand in sunlight you can feel the heat and see the light radiating from the Sun. (It's those waves that make the ground look like it's moving.)

Waves Come in All Shapes and Sizes

Once it was discovered that light was an electromagnetic wave, a number of other scientists and experimenters found more types of electromagnetic *radiation*. They established the terminology that helped to define the properties of electromagnetic waves.

Here is a summary of some of the most significant discoveries:

Universal Constants

To **oscillate** simply means to swing or move regularly back and forth. An electric current varies regularly between maximum and minimum values. An oscilloscope is an instrument that can visually display on a fluorescent screen the waveform of an electrical current. Oh yes, and you can oscillate yourself when you can't make up your mind about doing something.

➤ Joseph Henry, an unsung American scientist, produced and detected electromagnetic waves when Maxwell was only eleven years old. He built an electric circuit in which an *oscillating* electric current produced electromagnetic radiation. The more rapid the oscillation the more radiation it produced.

➤ Heinrich Rudolf Hertz (1857–1894) helped to clarify the content of Maxwell's theories and formulas by showing through experimentation how oscillating electric currents produced electromagnetic radiation. By pushing the design of Henry's circuit, he was able to find both the *frequency* and *wavelength* of the electromagnetic wave. From this he was able to figure out the speed with which the waves traveled. He found that although the frequency and wavelength change, all electromagnetic waves travel at the same speed: the speed of light.

➤ Wilhelm Conrad Röntgen (1845–1923) discovered x-rays in 1895. While performing experiments with very high-frequency radiation, he found that these electromagnetic waves passed through many substances. With a leap of insight, he directed these rays at his wife's hand that he placed in front of a photographic plate. After developing the negative, he saw that he had a photograph of the skeletal structure of her hand. Within months, x-rays were used in medical diagnosis.

➤ Ernest Rutherford (1871–1937) and his students studied the rays that come out of naturally radioactive substances such as uranium, and discovered that there are three kinds, which he named alpha, beta, and gamma. In reality the alpha rays were really helium atoms without their electrons (although no one even knew that atoms had electrons yet), and the beta rays were electrons.

So there you have a brief list of the early discoveries that were made regarding some of the more common forms of electromagnetic radiation.

Universal Constants

Frequency describes the number of cycles per second that an electromagnetic wave or electric current oscillates. It is expressed in hertz. Our household current is usually 60 cycles per second, or 60 hertz. There is a direct relationship between frequency and wavelength. The shorter the wavelength, the higher the frequency.

Wavelength describes the physical length of an electromagnetic wave, from crest to crest, and is measured in meters. For example, low frequency signals have a wavelength of 10,000 meters, or about 6 miles. Microwaves are about 10 centimeters, or about 4 inches. Shorter wavelengths are measured in angstroms. An angstrom is equal to one hundred millionth of a centimeter. X-rays are one angstrom long and gamma rays are 0.001 angstroms. The light we see has wavelengths between 4,000 and 7,000 angstroms.

A Little Bit of Quanta Goes a Long Way

At this point in the chapter, I want to cover how light was also discovered to behave as if it were a particle and not a wave. But this falls into the realm of quantum mechanics, which I haven't introduced you to yet. So for the time being I'm going to skip over some material that I'll come back to in the next chapter. The emphasis of this last section will be on explaining the dual nature of light.

There were a couple of problems in science that classical physics couldn't explain. These problems are what I'll explain in Chapter 11, because out of their solution came the birth of quantum physics. Within the context of these solutions it was discovered that not only could light behave as a wave, but also as a particle. As a matter of fact, all electromagnetic waves can also behave as particles. The name given to this is phenomenon *wave/particle duality*.

As you know, Newton firmly believed that light was made of particles; however, this view eventually changed, and everyone believed that light was a wave. Now, because of a few problems that classical physics couldn't explain, it was discovered by Max Planck and Albert Einstein that the energy in electromagnetic radiation comes in chunks of energy that Planck named *quanta*. Later, when referring specifically to light, these chunks of energy were called *photons*.

Light sometimes behaves like a wave and sometimes like a particle. Which way depends upon the circumstances, or to put it more strongly, how we want it to behave, or how we look at it. This sounds a bit strange, but in the quantum world, weird and unusual are normal. Let's revisit Thomas Young's two-slit experiment and update it to one of the most famous experiments in quantum physics.

Universal Constants

One of quantum theory's most revolutionary ideas is that all the constituents of matter and light are both wave-like and particle-like at the same time. This is known as **wave/particle duality.** Neither aspect is more primary than the other. The two compliment each other, and both are necessary for any full description of what matter and light really are.

We'll set up the experiment just like before. A light source placed in front of a barrier with two slits in it with a screen behind the barrier. The screen will act as our wave detector, which will enable us to know when light is behaving as a wave. We'll also add two particle detectors, one next to each slit on just the other side of the barrier, so when we turn them on we can measure each photon as it clicks past the detector.

In the original experiment the light source was shown through the two slits and an interference pattern was created on the screen, which meant that the light beam was behaving as a wave. And if we do that now, sure enough we get the same result. However, if we turn on the two detectors to measure the photons individually, we'll notice that they travel through only one slit at a time, not both, and they create a blob of light, rather than an interference pattern. When we shut the detectors off, the photons travel through both slits and create the wave pattern of interference on the screen. Who told them when the detectors were turned on or off? Why do they behave one way when they are counted individually and another way when they act collectively?

Universal Constants

Quanta, or the plural, quantum, comes from the Latin word for "how much." It describes the discontinuous or intermittent way in which electromagnetic energy radiates under certain conditions.

Many people think that Einstein named a particle of light a **photon,** but he didn't. Gilbert Lewis, a chemist, named the particle in 1926. Photons, or particles of light, are one form in which light can manifest, the other form, of course, is a wave.

You could say that when a particle is looked for, a particle is found. And when we look for a wave, a wave is found. Somehow there seems to be a direct relationship between how the experiment is set up and outcomes looked for. It's not really possible to say that light is really a wave that sometimes acts like a particle, or vice versa. The nature of light is a deeper and richer phenomena than either of these partial realities. Which side of its dual potential nature it decides to show depends entirely upon the experimental context in which it finds itself. We can never observe light outside of some context.

The potential for light to behave as both a wave and a particle at any given time was a totally new concept in physics and would be the first of many weird characteristics that would be discovered about the structure and behavior of the microcosmic world of quantum physics. And interestingly enough, the opposite paradox arose with solid matter, the objects we interact with in our everyday world. Matter is usually interpreted as consisting of particles, but they began to behave as though they had wavelengths. The small particles that make up solid matter were found to have wave characteristics. Even large objects like apples and ourselves have a wavelength. The wave nature of electrons is the physical basis of the electron microscope, which uses beams of electrons, whose wavelengths are millions of times shorter than those of photons, to view objects too tiny for examination under a light microscope.

Cosmonotes

The unusual ability of light to behave as both a wave and a particle is a central mystery that lies at the core of quantum physics. The very act of observing light in this context alters the outcome. And this is true of all subatomic particles. You can no longer separate the observer from the experiment. This realization would lead to many other principles and theories that would try to describe the paradoxical nature of the quantum world.

Our Journey So Far

I'd like to close this chapter by summarizing where we are in our cosmological journey and what's going to be coming up. We've more or less finished our historical survey of cosmology, at least through the beginning of the twentieth century. The two themes that we began with back in Chapter 1, "Once Upon a Time," the use of numerical symbolism and the observation of the heavens, have essentially become mathematics and astronomy. Mathematics is an integral part of the scientific method and is the fundamental tool and language of physics. And while we won't be using math to describe the cosmology of the microcosm and macrocosm, the mathematical theories will be translated into concepts and ideas that will help us understand the cosmos.

Cosmology's two main areas that I'll be discussing for almost the remainder of the book will be the microcosm and macrocosm. In the microcosmic universe we'll look at the weird world of quantum physics, explain the standard model predominant today, and move into supersymmetry and string theory. In the macrocosmic universe we'll explore Einstein's theory of general relativity and the two main theories on the origin of the universe. We'll also see how the individual study of both of these universes, the microcosm and the macrocosm, will eventually combine to provide a glimpse of a unified theory of cosmology. In the last couple of chapters, the final focus will be on the role that human consciousness plays trying to put it all together.

In the next chapter, we'll explore the birth of quantum physics, learn a few more things about the nature of light, and introduce you to relativity theory. You'll learn about thought experiments and frames of reference, and in the end you'll see that what you perceive, like what you believe, isn't necessarily the way things really are.

The Least You Need to Know

➤ Light was discovered to be an electromagnetic wave.

➤ Everyone believed that light waves needed to move in something, and that medium was called ether.

➤ The Michelson-Morley experiment proved to many scientists that ether didn't really exist.

➤ Michael Faraday developed the first theory of electric and magnetic fields.

➤ James Clerk Maxwell devised the mathematical equations to show that light traveled as an electromagnetic wave.

➤ Light has a dual nature—it has the potential to behave as a wave or as a particle.

Cracks in a Newtonian World

There is no other person who had as great an impact on twentieth-century cosmology as Albert Einstein. And while there are many well-known physicists such as Neils Bohr, Werner Heisenberg, Max Planck, Erwin Schrödinger, and more recently, Steven Weinberg and Stephen Hawking, they all stand on the shoulders of Einstein's discoveries.

Even now at the beginning of the twenty-first century, Einstein's theories continue to reveal new information about the universe, both at the microcosmic and macrocosmic levels. His revelations about the structure of energy, the principles under which gravity works, the nature of light, the relativity of space and time, and the proof of the existence of molecules would lead into important discoveries by other physicists and cosmologists. The state that cosmology is in today, which you'll get a chance to see later on, owes much of its success to the theories put forth by Albert.

The Clock Breaks!

To begin this chapter, I would like to summarize the basic assumptions about the dynamic structure of the universe that all scientists at the end of the nineteenth century generally accepted:

➤ All things moved in a continuous manner, and were subject to the *laws of motion*. These laws applied to all objects in the universe.

➤ Everything moved for a reason. These reasons were based upon earlier causes for motion. Therefore, all motion was determined and everything was predictable.

➤ All motion could be analyzed and broken down into its component parts. Each part played a role in the great machine called the universe and the complexity of this machine could be understood as the simple movement of its various parts.

➤ Space and time were considered to be absolute, meaning that they were fixed, never changing, and therefore universal constants.

➤ The observer observed and never disturbed. In other words, by watching an experiment you were distinct and didn't affect the outcome. Observation was not participation.

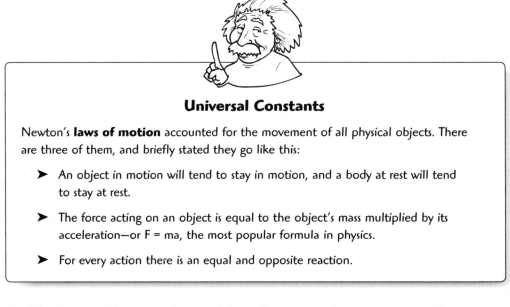

Universal Constants

Newton's **laws of motion** accounted for the movement of all physical objects. There are three of them, and briefly stated they go like this:

➤ An object in motion will tend to stay in motion, and a body at rest will tend to stay at rest.

➤ The force acting on an object is equal to the object's mass multiplied by its acceleration—or $F = ma$, the most popular formula in physics.

➤ For every action there is an equal and opposite reaction.

Within the next 50 years each one of these five assumptions was proven false. Einstein's special and general theories of relativity altered a few and the development of quantum mechanics changed the rest. There were two essential problems that classical physics couldn't seem to answer. The first one dealt with how heat was capable of producing light. And the second had to do with the constant speed of light. Let's take a look at each one of these, because out of these two problems quantum physics and relativity theory were born.

Hot Things Glow

Before light bulbs were frosted they were clear, and you could see the insides. The filament in a bulb was visible and it carried an electric current. As the current was increased, the filament would begin to glow and produce light. The color of that light would also change. The higher the current in the bulb, the hotter the filament got. And the hotter it got, the more the color changed. The question was, why? What was responsible for the changing color of the light? All hot objects, such as electrified light bulb filaments and heated branding irons, emit light. And if the light they emit is passed through a glass prism, that light will spread out into the familiar colored bands of the rainbow. These colored bands are called a *light spectrum*.

Sunlight produces a balanced spectrum of colors. There are equal amounts of all colors present. That is why sunlight appears "white" or colorless. All objects, no matter what their chemical nature and composition, send out light with the same color balance, if these objects are heated to the same temperature. It was the change in the balance of colors in the spectrum that was producing the different colors observed in the light bulb filament and a heated iron poker. The balance depended on how hot the glowing object became.

As an object is slowly heated to higher temperatures its characteristic color changes in a very predictable manner. Cool objects give off no apparent light. A hot poker glows red and at higher temperatures it starts to glow orange-yellow. At still higher temperatures, it glows blue. To get an idea of just what this looks like, look at a burning match. You'll see that the flame has different colors running through it. The temperature of the flame is not the same throughout, the bluer colors being the hottest part.

When the spectra from various objects are examined at different temperatures, you find that various colors are emitted in differing amounts. It is the shifting of these amounts of colors that changes the characteristic color of the glowing object. But as the object becomes hotter, its color becomes whiter and the spectrum becomes more balanced. Everyone believed that the connection between the temperature of a material and the color of light had to be a mechanical one. After all, the rest of the universe was believed to be just as mechanical.

Universal Constants

The **light spectrum** is a series of colored bands of diffracted light that are created when white light is sent through a prism. The colors are arranged in order of wavelength from infrared to ultraviolet. An easy way to remember the order of colors is by the acronym Roy G Biv: R = red, O = orange, Y = yellow, G = green, B = blue, I = indigo, V = violet.

After Maxwell's success in explaining that a light wave is an electromagnetic oscillation, scientists began to suspect that the different colors of light emitted in a heated object were caused by the different vibrational frequencies. So red light was thought of as having a lower vibrational rate or frequency than blue light. This led to the assumption that the light energy emitted by a glowing body should tend to be given off at a higher frequency rather than a lower one. This is based on an idea called light wave economics.

This idea simply expresses the relationship between the frequency of a wave and its length. Do you remember what that was? The higher the frequency of a wave, the shorter its length. And this related to a geometric factor that influences any glowing object. In other words, this geometric factor states that there are more ways for short waves to fit into any volume of space than for long waves. Light waves with very short wavelengths are able to take advantage of the space they find themselves in. And the geometric factor should produce short waves rather than long waves, or high frequencies rather than low ones.

What this meant was that heated objects should all produce light of very high frequencies, in the ultraviolet, x-ray, and Gamma range. There should be no change in color, only of intensity or brightness. But this obviously wasn't what everyone saw when objects got heated up. Things weren't their characteristic color and then all of a sudden became white hot. As long as light was considered to be a wave and thus mechanical, classical physics and Maxwell's mathematics could not explain light from heated bodies. Light energy would have to be considered to be something other than a wave. Max Planck and Albert Einstein would find the answer. You'll read all about their discoveries in Chapters 13, "Famous Equations and Pools of Jell-O," and 14, "That Old Quantum Theory."

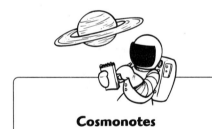

Cosmonotes

The inability for classical physics to explain the color change of heated objects is known as the "ultraviolet catastrophe." The problem is that if radiation is explained in terms of waves, using the same theory that describes sound waves, only the brightness of the light should change with temperature. The color should remain the same.

Faster Than a Speeding Light Wave

As previously mentioned, in Newtonian cosmology space and time were considered to be fixed entities. I use the word entity because both space and time are qualities of the universe whose existence, at least for us, is defined by these two terms. They are two things that are experienced by each of us all of the time, at least while we're awake. They form the very fabric of what we exist in, but defy clear definition. When we use words like holes in space, or tears in the fabric, even warping space—as you'll see when we get to general relativity—we have to ask, what exactly is being torn or warped?

Time is easy enough to measure, we have clocks for that purpose. But all they do is measure time. What is time? How fast is it going by? If time stops at the speed of light, as Einstein's theory of special relativity explains, (and which we'll get to in the next chapter) that means that it has to be moving by right now … from moment to moment. Well? Exactly how fast is it going by? As you can see, the *space-time continuum,* as it is referred to, has aspects of it that are conceptually difficult to grasp.

Universal Constants

The **space-time continuum** is made up of four dimensions: three of space—length, width, and depth, and the fourth, time. It allows any event in our world and the universe to be located. We are normally aware of moving through space, but don't often consider time as one of the dimensions we operate in—it's more of a commodity or separate thing. However, if you want to meet someone for lunch or see a movie, this involves a location in space as well as time. You really can't separate one from the other.

Unchanging Time and Space

Classical physics regarded space as a continuous expanse, extending in all directions. And as previously discussed, was found not to contain the ether that electromagnetic waves moved in. That was still a big puzzle as long as light was considered to be a wave—if it wasn't, then no big deal. Once light was shown to also behave as a particle, there was no need for any medium like ether for it to move in. But in any case, space was still considered to be unchangeable. And for now we'll leave it at that.

Time was also considered to be an unchanging thing. Regardless of how time may seem to change for us internally, outside of us clocks tick away time at a constant and even pace. But all of this was about to change. The one thing that scientists weren't absolutely sure of was whether or not light

Cosmonotes

It's important to point out that the speed of light is constant at 670 million miles per hour in a vacuum, like outer space. The speed of light does vary when traveling through different mediums such as air or water. But the decrease in speed is negligible so, for our purposes and discussion, the speed of light is considered to be a constant.

always traveled at the same speed. Maybe it slowed down or sped up at times. As it turned out, the only constant in the universe was the speed of light. Let's look at why this is so.

A good way to understand the constant speed of light is to use one of Einstein's favorite tools, a thought experiment. It's exam time at Star Fleet Academy, and for your final exam you and a friend (let's call him Al) have to set up an experiment to find out if the speed of light is constant, regardless of its source. You both go to the nearest space station, and while you remain on the station, Al rents a shuttlecraft from Stan's Cosmic Wrecks. You're making the final adjustments to very accurate measuring instruments, and Al is on the shuttlecraft setting up an ancient projectile weapon from the twentieth century.

Al takes off in the shuttlecraft and flies past the space station. You measure his speed as 50,000 mph. He then stops right out in front of you so that he's not moving, and he fires his high velocity cannon. You measure the speed of the cannonball at 10,000 mph. Now he flies by again, and fires the cannon while he flies by. You measure the speed of the cannonball and find that it's traveling at 60,000 mph. Where does that speed of the cannonball come from? It's equal to the combined speed of the shuttlecraft added to the speed of the cannonball. Okay, now in setting up the next experiment everything is the same except that you've replaced the ancient projectile weapon with a laser gun.

Passing Your Exams!

You already know the speed of the shuttlecraft but you want to be sure the instruments are calibrated correctly to measure the speed of light that is emitted from the laser gun. So Al stops in front of the space station and fires the laser gun. Sure enough the speed of light is 670,000,000 mph. He then flies by and shoots the laser gun. Based on the results of the first experiment you would expect that the speed of light from the laser gun should be the combined speed of the shuttlecraft plus the speed of light, or 670,050,000 mph. But to your amazement, the speed of light measured from your instruments is just 670,000,000 mph. The speed of the shuttlecraft has not added any more speed to the light emitted from the laser gun. You turn your final exam results into Star Fleet Academy and pass with an A. You've shown that the speed of light is constant regardless of its source.

It doesn't matter whether you are moving toward a source of light or receding from the source. Your movement has no impact on the speed at which light travels from its source. In Newtonian physics this doesn't make sense. It even goes against common sense. If you chase after someone running three miles an hour and you're running ten miles an hour, you'll catch up to him or her. If you take off after a light wave you would think you should be able to catch up to it. You would think that the light wave would appear stationary; the light wave would stand still. But there is no such thing as a stationary light wave. This thought intrigued young Albert Einstein. So much so that he spent 10 years of his life trying to understand the nature of light and its impact on space and time.

Mindwarps

A number of experiments have been conducted to prove that the speed of light is constant. One test was conducted in 1913 when the speed of light received from a stationary star was compared to the light received from a moving star. In both instances the speed of light was the same. Another test was made in 1955. In this case the speed of light was measured from the opposite sides of a rotating sun. While one side is turning toward you the other side is turning away. You would expect the light from the edge coming toward you would be faster than the speed of light from the edge turning away from you. Nope, light has the same velocity from both edges.

Cosmonotes

An important clarification to be made about any discussion of relative motion is that the type of motion we're talking about is constant velocity motion. That means that whatever speed you're traveling at, that speed remains constant. It's also called force-free motion. Because as soon as you introduce a force such as acceleration, or some kind of push or pull, that can be felt and you now know that something has changed. You become aware of movement.

It All Depends on Your Position

Copernicus's theory was the first hint that perhaps the nature of reality depended on the position of the observer. Galileo thought this to be true as well and developed the first ideas of relative motion. In fact, he proposed the idea that we could not tell by experiment alone whether we were moving or not. Here's his basic idea: If you were riding in the closed cabin of a steadily moving ship, without looking out of the porthole, you would not be able to tell whether you were moving or not. Allow your pet parrot to fly around the room outside of his cage. Watch fish swim around in their bowl. Toss objects back and forth. Notice how things fall to the ground. No matter how many experiments you perform, Galileo said, "You shall not be able to discern the least alteration in all the forenamed effects, nor can you gather by any of them whether the ship moves or stands still."

So it looks as though you can't tell the difference between standing still or moving if you don't have reference to something outside of you. You may be moving relative to the world outside of the ship, but not moving to the world inside it. Motion is relative because it depends on your point of view.

Let's say you ask your friend who is standing on the dock to tell you whether or not you're moving as you sail by in your ship. Your friend looks at you strangely and says, "Well duh, of course you are moving, can't you see the dock moving past you?" And of course you reply, "But how do I know that I'm moving and not the dock?" And then your friend says, "Or you could say that you are moving relative to my dock, but as long as you stand still in your cabin, you are not moving relative to your ship." And then you answer with the aura of wisdom that permeates all those who read *The Complete Idiot's Guide* books, "Cool, so I guess everything is relative."

This question would be a little more difficult for an outside observer to answer if you wanted to know whether the earth was moving or not. Where would you put the dock for your friend to stand on? How could you find a frame of reference where your friend would be standing still? The earth is whizzing around the sun at almost 20 miles per second. Our solar system is moving with respect to the center of our galaxy at 150 miles per second, and our galaxy—the Milky Way—is rushing toward our neighboring Andromeda galaxy (at least from the point of view of the Andromeda galaxy) at 50 miles per second. And if you look at the earth from a far-off quasar, you might see us speeding away at 165,000 miles per second, which is close to the speed of light.

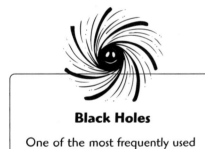

Black Holes

One of the most frequently used clichés about Einstein's theory of relativity is that it shows that everything is relative. The statement that, "everything is relative" really has no meaning. It's like saying everything is bigger. And if you ask, bigger than what? Apply that same question to everything being relative. In order to be relative, you have to be relative to something.

As you can see, the earth moves relative to the sun, our solar system moves relative to the galaxy, and the galaxy moves relative to the rest of the universe, and by current theories the universe is moving, too. But what is it relative, too? Other universes? Parallel universes? God? It goes on and on.

If you don't quite have the concept of relativity down yet, here's one more thought experiment that may help clarify it for you. Imagine that you're floating in outer space. There's no stars or planets or space ship for you to get your bearings or to give you a frame of reference. You just have a red distress light blinking on and off on your backpack. How you got in this situation is unknown to you. Off in the distance you see a yellow blinking light coming toward you. As it comes closer you recognize a friend of yours who seems to be in the same predicament that you are. Unfortunately, because neither one of you has a jet pack you can't maneuver toward each other. So you wave to you friend as she drifts by.

The unusual thing about watching your friend float by is that from her perspective she's not moving, you are. Let's turn it around. Your friend is floating in space and sees a red blinking light off in the distance coming closer. She now recognizes that it's you and waves to you as you go drifting by. From her point of view you are moving toward her. Each of you thinks that you are stationary while the other is moving. Each perspective is correct. And that is the fundamental principle of relativity. Motion is relative.

Cosmonotes

You may wonder why examples like the speed of light or relativity take place in outer space. It's a great place to explain how some theories work because there are two things that impact examples on earth. In space there is a vacuum, so objects move with no resistance and exhibit Newton's laws of motion perfectly. That's one of the reasons why satellites stay in orbit for as long as they do. There is very little to slow them down. The second is that there is no gravity to act as a force on anything. At least not in the regions of space the examples take place. There are some more examples coming up over the next few chapters that will also take place in outer space. It's just a great place to conduct thought experiments. Providing you have enough oxygen!

Sometimes Relativity Is Absolute

In this last section of the chapter I want to close with some thoughts on the impact Einstein had on classical physics and prepare the ground for his special theory of relativity. That's coming up in the next chapter.

Einstein extended Galileo's concept of relativity from motion to also include space and time. He looked inside the cabin of Galileo's moving ship (except that for him it was a light beam) and was deeply impressed with the fact that the laws of nature stayed exactly the same. Jump up and gravity pulls you down in the same way whether or not you're standing still or moving. Clocks tick, water flows, raindrops fall, and electricity works the same whether you are moving or at rest. Relativity established that the laws of nature are, in a sense, absolute and do not depend on the motion of the system. It's because they are absolute that you can't tell whether you are moving or not.

But this is only true in the everyday conventional world. What Einstein discovered was that the usual concepts of physics embodied in Newton's laws simply don't work at very high speeds or under conditions of extreme gravity or in many other situations. Newton's laws do not hold true in all frames of references, so the laws of nature depend on whether you are moving and what system you are in. The laws of nature depend on your point of view. This may seem beyond the realm of possibility, but remember what happened when we were looking at the dual nature of light? When the scientist looked for a particle, they found a particle. And when they looked for a wave, they found a wave. The very act of observation changes the outcome of the experiment, so maybe it does really depend on your point of view.

Galileo and Newton knew that motion was relative, but insisted that space and time were absolute. Einstein saw that space and time, and energy, and mass as well were also relative. However, all of this was true only because of the absolute nature of other universal constants, among them the absolute speed of light. That speed, as you know based on your Star Fleet Academy final exam, is an absolute 186,000 miles per second (or 670,000,000 mph) from any point of view, from any frame of reference, moving or not. The only reason "everything is relative" is that the speed of light and the laws of nature are not. And since light itself is really nothing more than the motion of magnetic and electric fields relative to each other, forces which are behind everything from the structure of matter to all of our processes including perception, Einstein clearly found a very fundamental "absolute" frame in which to construct his relative universe. And with that, it's time to look at just how special the special theory of relativity is.

Mindwarps

It's almost an oxymoron to label relativity as an absolute since the meaning of each is the total opposite of the other. The essence of absolutism in science is defined by a quality that is fixed, unchanging, complete, and whole. While at the heart of relativity is change and differing interpretations. But sometimes the best way to explain a concept is to put it in terms that describe it, paradoxically, by its opposite. This is a method that will be used more than once when we enter into the quantum universe.

The Least You Need to Know

➤ Classical physics couldn't explain two things: why a heated, glowing object gave off light, and the constant speed of light, regardless of its source.

➤ We live and move in an interconnected continuum of space and time, and you can't separate one from the other.

➤ Light is a universal constant; it always travels at the speed of 186,000 miles per second, regardless of its source or the frame of reference.

➤ The fundamental principle of relativity is that motion is relative.

Chapter 12

The Relative Nature of Space and Time

> ### In This Chapter
>
> ➤ Different kinds of time
>
> ➤ Simultaneous events
>
> ➤ Time dilation
>
> ➤ Motion's effect on space
>
> ➤ Movement through the space-time continuum

Einstein's theory of special relativity is based on two fundamental principles. They were both covered in the last chapter. Can you identify what they are? Here's a hint. One has to do with light and the other with motion. Tick, tick, tick … time's up. Without saying what they are yet, I can tell you that they are both intimately connected with time. That was also touched upon in the last chapter, but we'll go into it in much more detail now.

Okay, I'm sure you know what the two principles are, but just in case, let's go over them. Light, it has been determined, always travels at a constant speed, regardless of its source or the frame of reference. The other is the principle of relativity, which basically states that motion is relative. The combination of these two principles show that time is also relative. And as previously discussed, you can't separate time from space. So if time is relative, space must be also. How is it that time is relative? You'll know that by the end of this chapter.

It's About Time

Isaac Newton wrote in his *Principia:*

> Absolute space, in its own nature, without relation to anything external, remains always similar and immovable. Absolute, true, and mathematical time, is itself, and from its own nature, flows equably without relation to anything external.

This concept of time seems natural enough. We all accept it almost unthinkingly. But is the universe only a great big clock? Is time not related to anything else but itself? Every idea of time that we can possibly think of is deeply connected to a concrete physical event: the swing of a pendulum, the vibrations of a quartz crystal, the orbit of the earth, the quantum leaping of atoms, the motions of magnetic and electric fields, the lives of suns, your lunch with your friend, and on and on. Without such events, what would time consist of? You can't have time in a void because there would be nothing to relate it to. Time makes sense only when it's connected to something.

Here's another thought experiment. Imagine that a rock is the only inhabitant of the universe, and ask yourself these questions:

➤ What position does it have? Does this question have any meaning? No, because position can only be defined with respect to another position or thing.

➤ How about what size is it? Again we have nothing to compare it to.

➤ Does it continue to exist? Well, it doesn't change, there's nothing that's going to come along and impact it.

➤ So how can we tell the passage of time? You can't. Both space and time have no meaning at all.

That's a rather extreme scenario but one that illustrates the fact that it's the interaction of things, objects, people—all of the things that exist in the universe that provides a framework in which time and space have meaning. And, of course, we must also ask if time has any meaning for anything except us. Is time meaningful to animals, plants, or gods?

Time, most importantly, is a form of perception. We often speak of our "sense" of time. Just as there is no such thing as color without the eye to discern it, so an instant or an hour or a day is nothing without an event to mark it. We go about marking and discerning in a variety of ways. Perceptions of time change from country to country, from person to person, and even within us, from time to time. "A watched pot never boils" is a statement about the relativity of subjective time. Or "time's fun when you're having flies," oops, I mean "time flies when you're having fun" is another. Because we measure time by events that mark it, it should not be surprising that our sense of time is intimately influenced by the nature of the events themselves.

Mindwarps

Immanuel Kant, one of the all-time great philosophers, thought that the mind arranges and orders perceptions of the world through active organizing principles. These principles he called categories. Space, time, causality, and others are the categories that present us with the way in which we experience the world. He thought, as did many others, such as St. Augustine, that time was only a subjective experience that we projected out onto the phenomenal world. It has no meaning or reality outside of us.

Time also changes with perspective. This can be best explained by looking at the different ways in which time is categorized. Here's a list of just some of the modes of expression for time:

➤ Atomic time

➤ Geological time

➤ Biological time

➤ Astronomical time

Let's see how different each one of these are.

Atomic Time

The world of subatomic particles expresses extremely short, precise periods of time. The *atomic year* of hydrogen, for example, is 10 to the minus sixteenth power (that's one divided by the number 10 with 16 zeroes after it). This is infinitely short compared to an earth year. Yet the atomic year is incredibly long compared to the life span of many nuclear particles, which are millions of times shorter. The time scale at this microcosmic level is often counted in *nanoseconds*. And another area where time is this short is inside of a computer. For example, it may take a signal 12 nanoseconds to get from one place to another, which in some cases is a long time, but very short compared to snapping your finger.

Universal Constants

An **atomic year** is the time it would take an electron to "orbit" the nucleus of an atom if it were a miniature solar system. Different elements have different atomic years. In reality, electrons don't orbit the nucleus, but it's a convenient way to express a length of time for comparison.

A **nanosecond** is equivalent to one billionth of a second.

Radioactivity is another form of atomic time. There are certain forms of atoms that are unstable and after a while decay into more stable forms. For example, the nucleus of a carbon atom contains six protons and six neutrons and is known as carbon-12.

There is also another form or *isotope* of carbon, carbon-14, whose nucleus contains six protons and eight neutrons. During the process of radioactive decay, one of the extra neutrons emits a negatively charged electron and becomes a positively charged proton. Presto chango, abracadabra, the unstable carbon nucleus had changed into a stable nitrogen nucleus with seven protons and seven neutrons. What's really cool about this is that every 5,700 years, exactly half of a given number of carbon-14 nuclei will decay into nitrogen-14 nuclei. And every year after that the amount will be half of the year before. In other words, if you started with one million carbon-14 atoms, 5,700 years later you would have 500,000 carbon-14 atoms. And 5,700 years after that you would have 250,000 carbon-14 atoms. That period of time is called the half-life of carbon-14 atoms.

Cosmonotes

Knowing the half-lives of different isotopes is very useful because it can be used as a very accurate measuring tool for determining the age of different objects. It is especially useful in archeology, anthropology, paleontology, and other natural sciences. The reason is that all life on earth is carbon-based. This means that anything that was ever alive contains carbon atoms. Carbon-14 dating can be used to date the age of anything that was made from organic material, including man-made objects, and anything else that was at one time alive.

Geological Time

In geological time, an instant can be 10 million years because that amount of time is just $\frac{1}{450}$ of the Earth's history. A thousand years is an interval so short that it has little meaning to geologists and is only a passing moment. Geology breaks down the timeframe on earth based on principle physical and biological features. The four eras, which are the largest spans of time, covering hundreds of millions of years, are as follows:

➤ **Precambrian**   4,500,000,000 to 600,000,000 years ago
➤ **Paleozoic**   600,000,000 to 280,000,000 years ago
➤ **Mesozoic**   280,000,000 to 135,000,000 years ago
➤ **Cenozoic**   135,000,000 to 12,000 years ago

Each of these eras is broken down into periods or systems, and each period is further divided into epochs or series. The divisions are based on changes in the Earth, formation of landmasses, glaciations, and the development of different species and the extinction of others. It's not that dissimilar from biology in the way that everything is classified and categorized according to certain chief features.

Mindwarps

In traditional Chinese medicine, energy flows in the body through channels called organ meridians, related to 12 specific organs in the body. These meridians are the ones used in acupuncture and acupressure to release tension and stress, as well as to keep the body healthy. They have peaks of time when they flow according to the 24 hour clock—each one having a 2-hour time period. It's not that one stops and the other begins to flow, they're flowing all the time, but one takes the lead for that 2-hour period. It is similar to how birds fly in a V-formation so that the lead bird is supported by the rest of the flock and can relinquish the lead when it gets tired. Often, acupuncturists will put needle points on these meridians at their peak time to ensure the best routing of energy to the body and its related organ.

Biological Time

You may have heard some women refer to their biological clock ticking away. This usually means that they are ready to have children or that time is running out to have children. This analogy can be applied to other things in nature, too. (I'm sure nature doesn't have that on its mind.) There are all kinds of biological clocks. All living things need to tell time to survive, to coordinate their internal functions with the clocks of the outside world, to know when to hibernate, fly south, sprout, shed, or grow a winter coat. Our hearts need to know when to pump blood and our lungs when to breathe. Different organs in the same body may keep different times to different kinds of clocks, releasing chemicals in concert with communications from a central brain. Stomachs, livers, and sleep centers, may all tick to different, yet coordinated, times.

The limits of our ability to perceive time intervals even determine how we see the world. If we could sense intervals shorter than $1/24$ of a second, which we can't, you would see the dark gaps between the frames of a movie. If we could perceive much longer intervals of time, on the other hand, we could actually be able to watch plants or children grow.

Our experience of time definitely seems to change as we grow older. Some people think that this quickening sense of time depends upon the diminishing percentage of a lifetime that each hour or year takes up. To a year-old baby, a year is a lifetime—an eternity. To a 10-year-old, a year is but a tenth of his or her lifetime, and each hour is proportionately shorter. When we reach 50, time is passing five times faster still, and a year is only 2 percent of our lives. And if we reach 100, a year is only 1 percent. (If someone could come up with a pill to slow down time, it would probably outsell Viagra.)

Mindwarps

The hour is actually a recent addition to keeping track of time. Until the fourteenth century, days were divided into much less regular intervals of morningtide, noontide, and eventide. The first hours were flexible, they varied from summer to winter, and from daylight to darkness. Each day (dawn to dusk) and night (dusk to dawn) was divided into twelve equal parts. This meant that the hours of a summer day lasted longer than the hours of the same summer night. Winter daylight hours were correspondingly shorter and winter night hours correspondingly shorter. But even regular hours wouldn't work when the Industrial Revolution made it necessary for the trains to run on time and the workers to arrive for the five o'clock shift. Minute hands were needed to provide a more precise time interval, thereby introducing the need to be more accountable for one's time.

Astronomical Time

The time interval we call a year is marked by a single revolution of the Earth around the Sun. Our day is a single spin of the Earth around its axis and, long before we showed up on Earth, the month was very likely matched by the orbit of the Moon around the Earth. It doesn't any longer because the Moon's orbit is continually changing as the Moon moves further away. None of these astronomical measures could be considered absolute because they are constantly changing. Some 500 million years ago, our day was only 20 and a half hours long.

Let's leave our earth perspective for a moment and consider the concept of time on the planet Mercury. There the day is longer than the year. In other words, it takes Mercury longer to revolve on its axis than it takes for it to orbit once around the Sun. Talk about a confusing birthday party! We get so accustomed to thinking of days as the natural division of years into 365 parts that it's easy to forget that the day, like the year, just happens to define time relative to our planet's unique position in relationship to the Sun.

An important point to discuss before I finish this section on astronomical time is the concept of distance. In mathematics, there is a formula that defines the relationship between distance, speed (or rate), and time. Distance equals rate times time ($d = r \times t$). Anytime you move from one place to another, a certain amount of time expires. For example, I can walk from my house to the library. If I walk at three miles per hour and it takes me two hours to get there, I've covered a distance of six miles. So distance is the product of my rate, or the speed at which I travel, times the duration of time it takes for me to get from one place to another. This is all directly related to space and time through the absolute speed of light. Let me explain.

In order for us to locate ourselves in the universe in relation to all of the other billions of celestial bodies around us, we need to know how far away they are. Powerful astronomical telescopes and *radio telescopes* can locate planets, stars, black holes, quasars, and other galaxies. By locating their point of origin, or source of light, and other electromagnetic radiation, the distance can be calculated by how long it takes for the light to reach us. The speed of light is the measuring tool used to find out how far away these celestial bodies are. For example, it takes the light from the sun approximately eight minutes to reach us on earth. The light from some of our neighboring planets can take weeks and months to get to us. After that, distances and time are measured in light years or the distance that light travels in one of our earth years. That's pretty far away. Our closest star, Proxima Centauri, is four light years away, or 24,000,000,000,000 (twenty-four trillion) miles.

Time gets lost in the vast distance of space and, before we get lost in time, let's finally look at special relativity.

How Can You Be in Two Places at Once?

Now that we've spent a good amount of time talking about all the different kinds of time, let's get to the bottom of it all. If the nature of time hasn't appeared to be relative to you yet, then by doing another thought experiment I think you'll discover what Einstein did—time, like motion, is relative. For this experiment, we won't be going to outer space. Instead we'll just use a train and some mirrors to show that what appears to be a simultaneous event actually reveals that time flows differently for each of us.

Universal Constants

Many objects in the universe send out radio waves, and a **radio telescope** can be used to detect them. A large, curved, metal dish collects the radio waves and reflects them to a focus point above the center of the dish, rather as the curved mirror of a reflecting telescope gathers light waves from space. By detecting radio waves coming from galaxies and other objects in space, radio telescopes have discovered the existence of many previously unknown bodies, such as invisible giant stars that emit only x-rays and black holes.

We begin our thought experiment by measuring the distance between two poles alongside a railroad track and then find the midpoint between them. You are carrying a right angle mirror that allows you to see both poles when you stand at the midway point. Out of the sky two bolts of lightening strike the two poles simultaneously. You watched this occur in the mirror to verify that indeed the light struck both poles at the same time.

At this point, a train appears coming down the track with a friend of yours onboard who also has a right angle mirror. As the train arrives at the midway point, lightening strikes the two poles again. You observe the event as occurring simultaneously as before, but your friend doesn't. Why not?

Cosmonotes

To figure out the difference in time intervals for someone traveling close to the speed of light and someone stationary, Einstein developed a mathematical formula that he called the relativity factor. It adjusts for the difference in time between two frames of reference. For example, a person traveling in a spaceship at half the speed of light would have a 15 percent shorter time interval than someone at rest on earth. For every hour that goes by on board the spaceship, one hour and nine minutes goes by on earth. That's because 115 percent of 60 minutes is 69 minutes.

She sees the bolt hit the pole she's moving toward first and then sees the other bolt hit the pole behind her. The light has a shorter distance to travel from the pole that the train is moving toward than it does from the pole that the train is moving way from. Since the speed of light is constant regardless of the source, the only thing that can account for both of you seeing different events is that time must be relative. So a determination of whether or not two events are simultaneous depends on how you are moving as well. And since we are all in one way or another moving continually about through space, whatever happens to us in a different time also necessarily happens to us in a different space.

That simple thought experiment we just did lies at the core of Einstein's theory of special relativity. With time now understood as being relative, there was a significant result that had to be dealt with. What happens to time as we speed up the rate we're moving at? And if we travel as fast as light travels, what happens to time? And is time the only thing affected by light speed? What about space? To answer those questions we need to look at special relativity in a little more detail.

Keeping Track of Time

A common thought experiment in physics will be used to explain the slowing down of clocks. We're going to incorporate something called a light clock to illustrate this. This clock is made of two parallel mirrors with a photon bouncing back and forth between them. A round trip of the photon makes one tick. And let's say that there are a million ticks for every second. We also have a counter hooked up to the clock to show how many ticks have gone by. If I timed the dropping of a ball from the top of my house to the ground and the counter registered three million ticks, that means it took three seconds for this event to occur.

Now let's imagine that you're sitting at a table that has the light clock and it's just ticking away. On the other end of the table, a second light clock slides by at constant velocity. What will be revealed to us is that the moving clock will tick at a different rate than the stationary clock. How is this possible?

Well, let's look at the path that the photon must take in the sliding light clock from our stationary perspective. Now we're only going to observe one tick to illustrate the principle behind this experiment.

The photon starts from the base of the clock and moves to the upper mirror. But it must move at an angle as the clock slides by; otherwise it would bounce off into space. The photon bounces off the

Mindwarps

Einstein's famous theory of special relativity was described in a paper of his called *On the Electrodynamics of Moving Bodies*. He never called it the theory of relativity. That name was given to it years later by other scientists.

upper mirror and again travels a diagonal path to the lower mirror. Look at the figure below to get a clearer idea of how this works.

The motion of the photon in the sliding light as seen from a stationary perspective.

Now, remember that motion is relative, so from the point of view of the moving light clock, it is stationary and we are moving. That means that the photon is just traveling straight up and down from that perspective, even though from our point of view it is moving diagonally. Notice also that from our point of view, the distance that the photon travels is longer on the diagonal than within the up and down movement of the stationary clock. Since the speed of light is constant, regardless of motion, the distance the moving photon has to cover is longer than the stationary photon and therefore there will be less frequent ticks. The moving light clock will run slower! And guess what happens the faster it moves? The clock runs slower and slower until it stops altogether at the speed of light. There is no time at the speed of light. Time stops! Whether it's a light clock or any other type of mechanical, electric or atomic clock, the principle is the same.

Universal Constants

As an object approaches the speed of light, the time intervals, or the rate at which time passes, change. This is called **time dilation.** Time intervals are periods of time between events and they dilate or become longer as you approach the speed of light.

This explanation is what lies behind the many examples of *time dilation* found in introductory books on relativity. One of the classic stories is of the two twins. One stays home on earth while the other takes off on a journey through outer space. Many years pass. The twin on earth grows old, but the twin who was whizzing around the galaxy at close to light speed comes back still young. Time had slowed down for the space traveler from the point of view of his twin on earth and that is why he appears young. According to the space traveler, however, time has not slowed. The biological aging process remained the same, his clocks kept normal time and everything seemed "normal." Relativity is a property of time, not of clocks.

The Faster You Go, the Shorter You Get

You now know that time is affected by motion. And since time and space are inseparable, that means that space is affected by motion as well. Let's pay a visit to Casey and his cosmic locomotive. You ask him if he would participate in a science experiment, and of course, he couldn't be happier. You explain to him that you want to measure the change in the length of his locomotive as he shoots down the railroad tracks at close to light speed.

You get out your 100-foot tape measure and use it to find out how long his stationary train is. Okay, you did that. But now you realize that you can't measure its length while it's zooming past you, so you devise another method to do that. Remember that distance, or in this case the length of the locomotive, is equal to its speed times the elapsed time. You take out your stopwatch, which you always carry for experiments like this, and figure that if you start the watch the moment the tip of the train goes past you and stop it when the tail end goes by, you'll know the length.

As we learned in the last section, from Casey's perspective on board the train, he's stationary while you appear to be zooming by. That means that it will appear that your stopwatch is running slower too. This will result in a measurement of the locomotive that will be shorter, because if the watch is running slower, the elapsed time will be shorter, and consequently, when you do the math, the length of the train will be shorter than when it's at rest. Casey hops onboard, drives the train past you at 98 percent of light speed, and sure enough you not only perceive the train to be 80 percent shorter than at rest, but your stopwatch calculations verify this as well. Is your head zooming past you at light speed yet?

I have one last conceptual stretch for you before closing this chapter. We've discussed motion in relation to time and motion in relation to space. How about addressing the combination: motion through the space-time continuum? This gets a little tricky, but hang in there. The essential idea that I'll try to explain is one that Einstein believed to be at the heart of special relativity. And that is that an object's motion is shared through all four dimensions, the three of space and the fourth of time.

Mindwarps

Hendrik Lorentz (1853–1928) was a Dutch physicist who received the Nobel Prize in Physics in 1902 for his work on the theory of electromagnetism. His later work provided the bridge between Maxwell's work and Einstein's special theory of relativity. He developed the so-called Lorentz transformation equations, which describe the way space and time are distorted for objects traveling at a sizable fraction of the speed of light.

We all know what it's like to move through space—we do it all the time. But even if we're sitting still and not moving through space, we're still moving through time. When we set up a luncheon, we move through space to get there and arrive at an appointed moment "in time." I think this is already familiar, because of our previous discussion of time. So let's take this a step further. We know that when an object moves through space relative to us, its clock will run slower compared to ours. We saw that in the experiment with the light clock. This can be explained in another way as well.

If an object is stationary, relative to us, all of its motion is being used to travel through only one dimension: time. As soon as it begins to move, some of its motion is being diverted to moving through space, so there is no longer as much motion being used to travel through time, so time goes by slower, and that's why the clock of a moving object runs slower than one that is stationary, get it? The speed of an object through space reflects how much of its motion through time is being diverted. The faster it moves through space, the more its motion is being diverted away from moving through time, so time slows down.

Here's the final part. There is a limit to an object's maximum speed through space. And, of course, we know what that limit is. Right? It's the speed of light. There is nothing that can travel faster. That means that all of the object's motion is wrapped up in moving at the maximum speed through space. It has diverted its motion through time to accomplish this. Because there is no motion left for time ... there is no passage of time at the speed of light. The photons from the big bang are all still the same age today as when they first emerged.

That was some pretty heady stuff but great brain food nonetheless. There is still a bit of Einstein's theory of special relativity left, most importantly his most famous equation, $E = mc^2$. However, this equation deals with mass and energy and would be more appropriate with our discussion of Einstein's theory of general relativity. And that will be the focus of discussion in the next chapter.

The Least You Need to Know

➤ Time is a form of perception and is defined by the events that mark it.

➤ The relationship between distance, speed and time is expressed in the formula: Distance = Rate × Time.

➤ Einstein discovered that time, like motion, is relative.

➤ Time slows down as you approach the speed of light and is nonexistent at light speed.

➤ An object's motion is shared through all four dimensions of the space-time continuum—the three dimensions of space and the fourth one of time.

Part 4

From Here to Infinity

The ancient Greeks believed that we were the measure of all things. One of the interpretations of this was that we existed at the midpoint between the smallest and the largest things in the universe. To put it in today's terms, we would be halfway between the subatomic world of the microcosm and the billions of galaxies that permeate the macrocosm. That's an interesting thought when you consider that both ends of the scale have the potential to extend into infinity. Where's the halfway point on a string that's infinitely long?

Our study of cosmology has brought us to the halfway point as well. We began by looking at two fundamental themes, mathematics and observation. While we will continue to touch upon both of those in our discussion, at this point we're going to look at two other significant areas of cosmology. You're already acquainted with them somewhat, so in this part we'll go a little deeper and focus on examining what are called the standard models of the microcosm and the macrocosm. The strangeness of the quantum world will be explored and the leading theory of the origin of the universe, the big bang, will also be revealed. But before we do that, let's finish up Einstein's theory of special relativity and look at his reexamination of gravity.

Chapter 13

Famous Equations and Pools of Jell-O

In This Chapter

➤ The energy/matter connection

➤ Proper frames of reference

➤ The principle of equivalence

➤ Warped space and time

In the last chapter, you discovered that space and time were relative and that the only constant in the universe was the speed of light. After showing that these once thought absolutes were relative, Einstein extended that concept to show that energy and matter were not absolutes either. His most famous equation, $E = mc^2$, revealed the interrelationship between these two quantities.

We often hear reference to Einstein's theory of relativity. Well, in case you didn't know, there are really two theories. The first, the special theory of relativity, you've already read about in the last chapter and are about to finish up soon. The second, the general theory of relativity, deals with universal gravity. We'll be covering that in this chapter, too.

The Conversion of Energy

The conversion of matter into energy is an everyday occurrence. Every time you light a fire or burn coal, you are turning the energy of matter into the energy of heat. Imagine that before you built a fire you could weigh all of the molecules of air and wood that make the fire and, after the wood burned, you weighed the remaining air and the ashes. You would find that it weighed less. No, the missing weight didn't go up in smoke. It was transformed into heat energy, a precisely measurable amount of energy.

Einstein did a similar experiment only he used the *radioactive decay* (I discussed radioactivity in connection to carbon-14 in Chapter 12, "The Relative Nature of Space and Time") of the element radium instead of a fire. With radium, an alpha particle (or helium nucleus) is emitted during the decay process, and the loss of this particle converts radium into the element polonium. Having weighed the original amount of radium, he compared that to the total weight of the emitted alpha particle and the weight of the polonium. And as we knew he would find out, weight was lost. Let's put this into terms of Einstein's famous equation.

Universal Constants

The two main kinds of **radioactive decay** associated with atomic nuclei are alpha decay and beta decay. Remember those two particles from Chapter 10, "The Dual Nature of Light"? They have the effect of transforming a radioactive original nucleus (called the parent) into a nucleus of another element (called the daughter), which may or may not be radioactive itself. Simply put, it's a process in which an unstable nucleus or particle (like a neutron) spits out one or more particles and transforms into a stable nucleus or particle. Decay can involve the release of energy in the form of electromagnetic radiation as we noted in the discussion of Einstein's experiment.

$E = mc^2$

Albert knew the total energy released, designated by the letter E, and the missing weight (mass), represented by the letter m.

The key to putting it all together was the speed of light, c, times itself, or c^2. This is a huge number—34,596,000,000. These three things put together gives us his famous equation, $E = mc^2$. This translates as energy is equal to mass times the speed of light squared. This can be rewritten two other ways, $m = E \div c^2$ or $m \div E = c^2$. Notice that no matter which way it's written, regardless of the amount of mass and because the speed of light squared is such a big number, the amount of energy released is huge. That's why you get so much energy at the expense of so little mass and why nuclear explosions are so immense.

Infinite Mass

The interesting thing about mass is the degree to which it changes the faster it is accelerated. As you already know, as objects approach the speed of light, time slows down, length shortens, but mass increases. The conversion of energy into mass is not as familiar to us as the conversion of mass into energy is, but it happens just as often. Every time you run, you put on a little extra mass. (And here you thought that all that running would help you lose weight!) A tightly coiled spring has more mass than the same spring relaxed because there is extra mass due to the energy coiling puts into it. In other words, the faster an object moves the more mass it gains.

Mindwarps

If your mass is around 150 pounds, you contain enough energy of mass to power a small city for a week—that's if you could convert it. And this is one of the reasons why no one discovered that mass was convertible into energy before Einstein. It just didn't seem available. But finally, about 40 years after Einstein developed his famous equation, the first nuclear bomb converted the mass of a small amount of uranium into energy and revealed that it was, in fact, possible to do so.

Cosmonotes

There's an interesting distinction that should be made regarding weight and mass. In physics, the term usually used to denote the stuff that matter is composed of is mass. This is the sum total of all of the protons, electrons, and neutrons in the object. It's also defined by the quantity of matter in an object as measured in its relation to inertia, or the tendency to remain at rest, or if moving to keep moving. While weight is defined by the force of gravity acting on an object and is equal to the mass of the object times the acceleration of gravity, it can vary in different gravitational fields. You weigh less on the moon than on earth. Mass is normally expressed in metric terms, such as grams or kilograms. There are also different terms for mass depending on how it's used. There is gravitational mass, inertial mass, relativistic mass, and rest mass.

Well then, what about photons? They must weigh tons if they are traveling at the speed of light, right? Nope, photons are mass-less. They contain all of their mass in the form of motion energy. It is very similar to our discussion at the end of the last chapter. Let's look at the similarity.

Remember that the photon had diverted all of its motion through time to attain the maximum speed of motion through space. This is sort of the same scenario only instead of time it's mass. The photon has no mass and experiences no time because all of its energy of motion, the maximum speed that any object can travel through space, has been diverted away from the energy required to move through time and contained in its mass. The photon is pure energy, having no mass and experiencing no passage of time.

This energy/matter relationship is the reason behind the observable fact that light speed is the speed limit of the universe. No energy or information can travel faster than light because as anything begins to approach the speed, it gains an ever-increasing amount of mass. As previously mentioned, mass is a measure of inertia— a resistance to change in motion. So the more speed or motion something has, the harder it is to make it go faster, because it also has become more massive. Eventually, the object gets infinitely massive, which means that it would take an infinite amount of force to make it go any faster. So even inertia is not absolute. Inertia increases the faster you go.

Mindwarps

In giant subatomic particle accelerators, electrons pushed to 99.999 percent of the speed of light gain 40,000 times their original mass. And just to show the change in a few decimal places, a muon (a particle identical to an electron except it's 200 times heavier) taken to 99.99999999 percent becomes 70,000 times heavier than its original mass. These particles pushed to nearly light speed don't gain velocity as much as they gain another form of energy—mass. So in reality, these particle accelerators aren't in the business of accelerating particles to high speeds as they are in the business of building them up to more remarkable masses.

Recapping Relativity

We've now covered the relative nature of five things: motion, time, space, energy, and mass. Let's just summarize what we know about this before moving on to Einstein's theory of general relativity. To begin with, everything looks different from different points of view. If some people are moving past you close to the speed of light, they seem to get very massive. They however, will not feel themselves getting more massive. They will see you getting massive.

Objects, people, planets, and stars—anything that moves very fast in relation to you—appear to get more massive. The increase in mass is based on motion: the greater the speed, the greater the increase in mass. But motion, as Galileo observed, is relative. A clear perspective about who is moving and who is not depends on your frame of reference, your point of view. So a judgment about how massive something is must also depend on your point of view. And what holds for mass must also hold for energy. If the energy of motion at high speeds can be converted into matter and vice versa, then it's clear that how much energy you have depends on how much motion there is. So energy in this sense is relative, too.

Mindwarps

It's interesting to extend Einstein's idea of a "proper" view to other areas. From a cultural perspective it's often the case that whatever we do is proper and normal and whatever "they" do appears peculiar and weird. Have you ever met a parent who thought anyone but himself or herself knew the "proper" way of raising children? Even national populations consider their political ideas as normal and proper, while other nations aren't quite at the same level. You can tie this all into our previous discussion of truth from Chapter 9, "Evolution vs. Creationism."

This same understanding also applies to time and space. Travel at speeds nearing the speed of light cause your clock ticks to tick more slowly and your space to contract. But these effects are necessarily relative to something else. For example, if a friend of yours stays home while you whiz off on a quick trip around the galaxy, you may return to earth still relatively young (of course, that all depends on how old you are when you leave) years after your friend has died of old age. But you stay young only relative to your friend and the rest of your family at home. According to your internal biological clock, or wristwatch, or any other timekeeping apparatus, time flows as always. You can't tell that time has slowed down for you, because your entire frame of reference is running slower; your heart rate has slowed down, your breathing slower, all bodily functions are operating at a reduced pace. Even atomic clocks would run slower.

The result of all this is that you have no way of knowing whether you are more massive or not, whether your time has slowed or not, whether your space has contracted or not, or even if you are moving or not, just like the person on Galileo's ship. Your mass is normal mass, your time is normal time, and your space is normal space. Everything in your view seems appropriate and proper, while things and people moving relative to you appear slow moving, squashed, and weird. Einstein called the view which you see the "proper" view, to distinguish it from other frames of reference.

The bottom line to all of this is that things look different from different points of view. The simple observation of a cardboard box can look different depending on whether you're speeding past it in a car or standing next to it. The box itself doesn't change, and the laws of nature don't change. The relationships between objects and events don't change. That's why you can't tell whether or not you're moving, or if your clock is running slow. From within the confines of our own limited perspective, things seem very different where in truth they are very much the same. The equations of relativity can provide a kind of dictionary that helps to translate from one frame of reference into another. They allow you to move step by step to someone else's point of view. Relativity can essentially mean that everybody can agree on the facts of a situation, even though we all see it from vastly different points of view. So motion, time, space, energy, and matter, we could say, form the core cosmological structure of the universe within the framework of the scientific paradigm. And our understanding of these five, in essence, really seems to depend on our point of view.

Relativity theory also shows that the perspective of a constantly changing, dynamic cosmos has the potential to be understood from a *unified system*. And although the ideas contained within relativity theory have united these five things through changing perceptions, it's not quite that simple. There are other forces and structural problems that need to be addressed. We've only dug three feet out of a ten-foot hole. But unification is the ultimate goal. As we move through the remainder of the book, that will become a major focus. The next thing we have to examine is the universal force of gravity.

Universal Constants

One of the main goals of physics is to bring together all of the forces, energies, particles, and theories into a **unified system.** This is the Holy Grail that physicists are searching for. It's also known by various other names such as the Unified Field Theory, the Theory of Everything (TOE), or a Grand Unified Theory (GUT). Einstein tried to accomplish this but never succeeded. It's been partially completed, but the final unification of the microcosm with the macrocosm—quantum mechanics and general relativity—is still on the drawing board. The best theory to date is the string theory, which we'll be getting to in Part 5, "Supersymmetry, Superstrings, and Holograms."

Thank God, the Ground Broke My Fall!

Newton's law of universal gravitation works wonders. It allows us to send men to the moon and probes to Mars, know the position of any celestial body and a lot more. There's just one small problem. It doesn't work with the special theory of relativity. Einstein recognized this, which is why he developed his theory of general relativity.

We all know by now that there is nothing that can travel faster than the speed of light. This applies to information, signals, and influences as well. Newton's theory of gravitation explains that one body exerts a gravitational pull on another body with a strength determined only by the mass of the objects involved and the degree of their separation. (If you need to, go back to Chapter 8, "Physics 101," and review the definition.) What this means is that if their mass or separation should suddenly change, based on Newton's theory, they would immediately experience a change in their mutual gravitational pull.

Let's look at an example. If for some reason the Sun were to explode, normally it would take eight minutes for the impact of that event to reach us. That's because that's how fast it takes light to reach us. Under Newton's theory of gravity, the explosion would register instantaneously and the gravitational effect would be transmitted to earth with no time going by. Well, according to special relativity, that's just not possible.

Einstein recognized this conflict and set out to devise a new theory of gravity. Our understanding of time and space was about to go through yet another alteration.

Cosmonotes

A common question to ask is, "If there are these three or four problems that have been discovered in classical physics, why is it still used and taught?" Well, there is nothing inherently wrong with classical physics. Yes, there are problems that it can't deal with (extreme speeds, the quantum world, gravity), but in our everyday world it works just fine. Skyscrapers and bridges are built, the speed of baseballs is measured, and shuttles orbit the earth, using the physics of Newton and Maxwell.

Einstein's Happiest Thought

There are two main aspects of general relativity. The first, which I'll be discussing shortly, has to do with something called the principle of equivalence. The second aspect, which I'll get to as well, has to do with the shape of space. Einstein realized that he had to extend his special theory of relativity to resolve the conflict of "action at a distance," which was another phrase for the instantaneous transmission of gravity through space previously discussed. He also knew that his special theory didn't account for accelerated motion, only constant velocity motion. The following is a quote from Einstein revealing what he was about to discover:

> The theory of relativity resembles a building consisting of two separate stories, the special theory and the general theory. The special theory, on which the general theory rests, applies to all physical phenomena with the exception of gravitation, the general theory provides the law of gravitation and its relation to the other forces in nature.

So to summarize the main difference between the two we could say:

➤ The special theory deals with frames of reference that are moving in a straight line, toward or away from each other, and are not accelerating, but instead are moving at constant or uniform speed.

➤ The general theory defines the relationships between objects throughout all of space, moving in any direction, with or without acceleration.

Out of his search to find the extension of his special theory, Einstein one day had what he considered "the happiest thought of my life." This thought was the realization that without a frame of reference you would not be able to tell the difference between gravity and accelerated motion. Let's do a few thought experiments so you can experience his happiest thought for yourself.

Let's begin with a ride in our rocket ship to a distant planet. The ship is accelerating to achieve top speed. All of a sudden you drop your cell phone. Now if you were simply drifting around in orbit, your cell phone would also be floating next to you in space. But because the ship is accelerating, the floor soon overtakes the cell phone and it appears to "fall" to the floor. From the point of view of someone, let's say on a passing planet, they would see the cell phone remain stationary while the accelerating ship moved to catch up to it. (Yes, you have a transparent ship so all the extraterrestrials can see everything that you do.) But to you inside the ship, it would appear as though some outside force, like gravity, was attracting the cell phone to the floor. Get the idea?

Here's another example. Imagine that you're on a return journey from a distant planet to earth. The controls are on automatic pilot and the ship is accelerating to get home in a hurry. You fall asleep, having complete trust in the ship's ability to get you home and even land without your help. You wake up and can't tell if you landed on earth or if you're still in outer space. You bounce a couple balls, pour yourself a drink, and get ready to disembark, because it seems that you've landed. However, just before you open the door, you happen to look out the portal window and see that you're still flying through space. As before, inside Galileo's ship, you have no way of knowing what is influencing your perspective. In this case it's gravity; in Galileo's ship it was motion. This was Einstein's happiest thought. He had the realization that you can't tell the difference between gravity and accelerated motion without a frame of reference. This is known as the *principle of equivalence*.

Gravity Is Relative, Too

Equivalence can be easy to get used to. When astronauts accelerate from the launching pad in a

Universal Constants

The **principle of equivalence** simply states that you can't tell the difference between gravity and accelerated motion without a frame of reference. Gravity is equivalent to acceleration. It is one of the two main ideas found in Einstein's theory of general relativity.

Universal Constants

Centrifugal force is what you feel on a spinning amusement park ride. It's what keeps you pinned to the sides of that ride whose floor drops out from underneath you, while you remain stuck to the inside of the rotating walls. You can also experience it whenever you spin an object on the end of a string over your head. The outward force is what keeps it spinning instead of hitting you in the head.

rocket toward outer space, they measure the force of acceleration in so many Gs. A "G" is the designation for the force of one unit of earth's gravity. Two Gs would equal twice the force of the Earth's gravity, and so on. The ideas we have about space stations orbiting the Earth substitute another kind of acceleration for gravity, *centrifugal force*. Huge circular rings of the space station swirl around in space, throwing people and objects outward like stones twirled on strings. If the ground is built on the outer edge of the circular space station, then the centrifugal acceleration will be exactly equivalent to gravity.

Mindwarps

Sometimes trying to understand general relativity isn't easy. It's a lot like the story of the engineer who tried to explain to a peasant how a steam engine works. After he described where the steam goes and how it moves through the engine and moves the piston, the peasant replied, "I understand all of that, but where's the horse?" (Output is measured in horsepower.) And that's how some people feel about general relativity. They get the details, but still don't know where the horse is.

Interesting enough, Newton believed that accelerating motion created by centrifugal force was absolute, not relative. He said that while steady motion was relative, as shown by Galileo, accelerating motion was not. If you spun a bucket of water, the centrifugal force would make the water rise at the sides, very much in the same way that the spin of the earth causes it to bulge outward at the equator. This bulge was clear evidence that these things were moving and not at rest. However, around 1900, Ernst Mach (one of Einstein's revered mentors) explained that if you spun the whole universe and kept the bucket or the Earth at rest, you still couldn't tell whether you were at rest or accelerating.

The idea that a force like gravity can be relative can be a difficult idea to get used to. When you push something to make it go, or throw a ball, there doesn't seem to be anything relative about it. If you push something with a large force and it goes farther than when you push it with a small force, there's nothing relative about that either. But if motions are relative and motions are the result of forces, then it's easy to see that forces have relative qualities as well.

Curved Space

The second aspect of general relativity is based on the exotic notion of space being something that can bend, warp, and ripple. Einstein's theory of general relativity describes the force of gravity as the unseen geometry of space. In other words, gravity is not a force that makes things fall but is rather a property of space. Things fall because that is their normal straight-line path in a curved, four-dimensional space-time continuum. But curved space is really a weird concept to grasp. Is there anything called straight space? A straight line is the shortest distance between two points and usually the shortest distance means line of sight or the path of a light beam. But we assume that light travels in a straight line. If the light-beam curves, is it going straight in curved space? Or is it curving in straight space? On some occasions the shortest distance between two points is the path that takes the longest according to your watch, because time slows as you go faster.

Well, without confusing you any further, let's do another thought experiment to give you an idea of how and why space curves. Some of the most common ways to think of space is to see it as an elastic medium. Think of a waterbed or a sheet of rubber or a pool of Jell-O. If you place a heavy object like a bowling ball on it, the mass of the ball depresses into the material. Now if you roll another smaller ball, like a marble, out onto your bed sheet or Jell-O, the marble will fall into a curved path made by the depression of the bowling ball and circle around it. If the marble has just the right speed and direction, it will go into orbit around the bowling ball.

Mindwarps

What exactly is space? No one really seems to know. Some philosophers and scientists think that space and time are just states of consciousness that we experience as human beings. The fact that space and time take on very different aspects when we sleep and dream suggests that they only have the qualities that we imbue them with when we're awake. On the other side of the coin is the notion that space itself is "alive" in a certain sense. The shape of space responds to objects in the environment. This is close to the ancient cosmological view of Plato and Pythagoras who thought that space was an organic whole. It could be just one of many dimensions that contain universes within universes. Someday all will be revealed, or maybe not.

Mindwarps

One of the most stunning implications of general relativity was the prediction in space of black holes. A black hole is a compressed star whose gravitational warping of space has turned in on itself. The massiveness, in terms of density of matter in these black holes, is so great that they severely warp space into a dark funnel. There is nothing that comes close to these dark funnels that can escape its gravitational pull, not even light. But it's unknown what happens at the deepest point of the vortex of these black holes.

If you replace the Jell-O or the bed sheet with space, the bowling ball with the Sun, and the marble with the Earth, you can get an idea of how the Earth orbits around the Sun. If you apply this model to the planets and moons in our solar system you can see how its fundamental structure is laid out. The only other thing you have to realize is this warping of space happens in all directions, not just the two-dimensional surface of our Jell-O pool or rubber bed sheet. Think of the bowling ball as being covered by the Jell-O or rubber sheet and depressing it in every direction.

Here's a list of some of the features that are a result of this aspect of general relativity:

➤ The more massive an object is the more it will warp the surrounding space.

➤ The more massive an object is the greater the gravitational influence it will exert on other bodies.

➤ The amount of spatial warping decreases as the distance from the massive object increases.

➤ All objects warp space, but the degree to which they do depends on their mass.

➤ Because a warp distorts space, it also must distort time. This occurs because you can't separate space from time.

➤ The more massive an object is, the more time will slow down the closer you get to it.

Einstein was able to develop this theory of a mass warping space by combining what he learned from special relativity and the principle of equivalence. Out of this came an answer to the question that Newton had asked. Newton believed that "gravity must be caused by an agent," although he had no idea as to what that agent was. Part of his alchemical quest, which we covered back in Chapter 7, "Cosmology in the Age of Enlightenment," was devoted to trying to find out what this agent was. Einstein revealed that the agent of gravity was the very fabric of space itself.

The Least You Need to Know

➤ Einstein's famous equation, $E = mc^2$, reveals that energy and matter can be converted into each other.

➤ Space, time, energy, mass, motion, and gravity are all relative.

➤ The principle of equivalence states that accelerated motion and gravity are equivalent with a frame of reference.

➤ Gravity is caused by massive objects that warp the surrounding space.

➤ Time and space are both affected by massive objects.

➤ The general theory of relativity consists of two main aspects, the principle of equivalence and the concept that gravitation is caused by the distortion of space.

That Old Quantum Theory

Einstein's two theories of relativity have shown us that when things move very fast or when objects get massive, the universe exhibits very strange properties. The same is also true of the microscopic world of quantum interactions. The deeper we delve into the macrocosm and the microcosm, the further we get away from the things that make sense to us in our everyday world.

At the heart of relativity theory is the idea that what you experience is real for you but not necessarily for someone else—it depends on your frame of reference. This can also be said of the quantum world, but with less certainty. Because what we find within the framework of quantum mechanics is that events are based on probability and not certainty. And the probable outcome of any event often depends on what you're looking for.

In Chapter 10, "The Dual Nature of Light," I discussed the wave/particle duality of light and touched on some aspects of the quantum. This chapter will extend that discussion by inquiring into just why energy has to come in chunks, or quanta, and how this would lead in the next chapter into the bizarre nature of quantum interactions where everything is so uncertain. We'll also look at the key theories of some of the great physicists who helped to define this weird quantum landscape. Next stop, the quantum zone.

Founding Fathers

There are many physicists who have contributed to the development of quantum mechanics. For our purposes we're going to look at the handful whose theories laid the foundation that others would build upon. Here's a list of those that we'll be taking a look at. If I omit any, it's not because their contributions weren't significant, it's only because I can only cover so much as an introduction to this huge body of material.

➤ Max Planck is often called the father of quantum mechanics. He coined the term quanta and developed the solution to the ultraviolet catastrophe (discussed in Chapter 11, "Cracks in a Newtonian World"). His main contributions were the theory that electromagnetic radiation happens in discrete quanta and the discovery that the size of each quantum is associated with a universal constant, a physical ration or proportion that stays the same in all circumstances and all frames of reference.

➤ Albert Einstein described the photoelectric effect that explains how electrons are released from a substance under the influence of light or other electromagnetic radiation. This was a key step toward the concept of wave/particle duality.

➤ Niels Bohr developed the theoretical model of the atom that most of us are familiar with. This model has electrons orbiting around the nucleus like planets orbiting around the Sun. He also promoted the Copenhagen interpretation of quantum mechanics.

➤ Louis de Broglie proposed that all material particles such as electrons could also be described in terms of waves. This led to the replacement of Bohr's model with one that explained the orbits of electrons as waves.

➤ Werner Heisenberg discovered his now famous uncertainty principle that simply places the role of uncertainty as a core characteristic of the quantum world.

➤ Erwin Schrödinger described quantum mechanics in terms of wave mechanics and had a famous cat named after him.

These six physicists along with three others—Wolfgang Pauli, Max Born, and Paul Dirac—were all together in 1927 at the *Solvay Conference*. It was during this conference that the essential concepts of quantum theory were formulated. What I will do for the rest of this chapter is to discuss each of the founding fathers listed above and explain some of the basic concepts of their contributions. Let's begin with a little overview of just what quantum mechanics is all about.

Universal Constants

A few years before the outbreak of World War I, the Belgian industrialist Ernest Solvay (1838–1922) sponsored the first of a series of international physics meetings in Brussels called the **Solvay Conference.** Attendance at these meetings was by special invitation, and participants, usually limited to around 30, were asked to concentrate on a pre-arranged topic. The first five meetings held between 1911 and 1927 chronicled the development of twentieth-century physics. The 1927 gathering was devoted to quantum theory and attended by nine of the most brilliant theoretical physicists. Each of the nine would eventually be awarded the Nobel Prize for their contributions.

Quantum Leaps

What is quantum mechanics? Well, essentially it's the mechanics of quantized things. Okay, so what does that mean? In physics, mechanics is not about fixing your car, but is the explanation of the way things work in terms of energy, forces, and motion. And anything that is quantized is simply expressed in multiples of some small measurable unit. As you already know, before the turn of the century, the way things worked was explained by Newtonian mechanics or classical physics. And the central feature of Newtonian mechanics was that everything was continuous; things flowed smoothly through space; energy could come in an infinite range of amounts; light undulated in a continuous wave; there was no minimum amount of anything.

Quantum mechanics changed all that. Now energy, light, force, and motion all came to be quantized. For something to be quantized, you can't have just any old amount; you can only have multiples of certain minimum quantities. Quantum mechanics meant that all the qualities of subatomic things, and by extension, all things, were precisely quantifiable. Nature revealed herself to be somewhat grainy or jerky, jumping from one quantum amount to the other, never traversing the area in between. This leads to an uneasy uncertainty about what is going on between those quantum states, or to put it into a popular phrase, *quantum leaps*. In fact, as it turns out, there is no way of knowing the exact state of things between quantum states. In our perception (and that is a key ingredient), there is no transition between quantum states. You can have one or two or three units of energy or momentum or light or force or matter or whatever, but there is no such thing as one and one-half or two and three-quarters units.

Universal Constants

A **quantum leap** is a discontinuous transition between quantum states. What this means is that an electron in one energy level in an atom jumps instantly into another energy level, emitting or absorbing energy as it does so. There is no in-between state, and it doesn't take any time for the leap to occur. The leaps also occur at random, selecting from the options available to the quantum entity in accordance with the strict rules of probability. So a quantum leap is a sudden change in a system that occurs on a very small scale and is made at random.

An example of this might help you understand. Imagine that a child is jumping up the stairs. A quantum leap is the transition from one state to the next state, the same as a child jumping from the first stair to the second stair. Depending on the amount of energy she has, she can also jump up to the third stair, the fourth, or the sixth. But she can't jump in between the stairs and land safely.

Everything in the quantum mechanical universe, which is the universe we live in, happens in quantum leaps. The uncertainty associated with these leaps will be explained when we get to Werner Heisenberg in Chapter 15, "Chunks of Uncertainty." The important point I want to mention is that all of this comes into acute focus when we actually set out to measure things as small as atoms and try to describe them in language of typical Newtonian systems. Lost? Don't worry, it will become quantumly clear shortly.

Planck's Constant

In Chapter 11, we briefly covered the ultraviolet catastrophe. If you don't remember exactly what that was, now would be a good time to go back and review it. Max Planck was the man who came up with the solution to it and gave birth to quantum mechanics. Let's look at what he discovered.

The ultraviolet catastrophe deals with something called *black body radiation*. The important point about this type of radiation is that its properties depend only on temperature. If we plot a curve on a graph representing the spectrum of the radiation, it looks like a smooth hill, with a peak at a frequency (color) that depends on its temperature. The hotter the radiation, the higher the frequency at which the peak occurs. This relates directly to the earlier explanation in Chapter 11 where the color of an object that was heated slowly changed color from red to white.

Universal Constants

A **black body** is a hypothetical object that absorbs all electromagnetic radiation that falls on it. Such an object, if heated, would be a perfect radiator, producing **black body radiation.** Such a hot black body would be no longer black, because it would be radiating visible light. The best example of a black body in a laboratory is a container with a small hole in it into which radiation shines and is trapped. When the container is heated, radiation bounces around inside of it and eventually escapes from the hole as black body radiation.

Until Max Planck developed the first quantum theory of radiation, the shape of the black body curve was a mystery, since it could not be explained by the classical physics of Maxwell's equation of electromagnetism. The problem was that if electromagnetic waves are treated mathematically in the same way as strings on a violin, and if waves can be any size, classical theory predicts that when energy (heat) is put into any object and radiated as electromagnetic waves, the amount of energy radiated at each frequency is proportional to the frequency. In other words, the higher the frequency, the more radiation there should be. A black body should emit huge amounts of energy in the highest frequency (shortest wavelength) part of the spectrum, in the ultraviolet and beyond, which of course it doesn't.

Planck found a way to avoid this problem. He cut up the radiation, mathematically, into chunks, or quanta. At a particular frequency f, each quantum of radiation has an energy E, given by the equation $E = hf$, where h is a constant in nature, now known as *Planck's constant.* Let's see how this works.

In any object, the energy is distributed among the atoms according to the temperature of the object. A few atoms have low energy, a few have high energy, and a lot have the middle amount of energy. This large amount of middle energy atoms increases as the temperature increases. Each atom can emit electromagnetic radiation. For very high frequencies (large values of f), the energy needed to emit one quantum of energy (e) is very large, and only a few atoms in the black body will have that much energy available, so only a few high-frequency quanta are radiated. At very low frequencies, it is easy for atoms to emit low-energy quanta, but they have so little energy that even added together they do not amount to much. In between the two extremes, however, there are many atoms that have enough energy to emit moderate-sized quanta of radiation. These add up to produce the peak in the black body curve. And the peak shifts to higher frequencies for hotter bodies, because in hotter bodies there are more individual atoms with greater amounts of energy. And that's it!

173

Universal Constants

Planck's constant is the ratio of a particle's energy to its frequency. Mathematically this is written $h = E \div f$, where h is the symbol for the constant. So, if a particle's frequency increases, its energy must also increase. If its frequency decreases, its energy will decrease as well. But Planck's constant always stays the same and is always equal to one quantum. Another way of putting it is that it relates energy's particle nature to its wave nature. It's represented by the number 0.000 000 000 000 000 000 000 000 006 626, or about a billionth of a billionth of a billionth of 1. But it is not 0. If it were, we wouldn't be able to sit in front of a fire. It's time to be grateful for the little things.

Photoelectric Effect Explained, the Quantum Strikes Again

Einstein was the first person to take seriously the idea of light quanta and to treat them as more than a mere mathematical trick used to explain the spectrum of black body radiation. Building on Planck's theory, Einstein was able in 1905 to explain another experimental puzzle, the *photoelectric effect,* by proposing that light is a stream of photon particles, each carrying one quantum of action. His theory that electrons are knocked out of a metal surface (such as zinc or copper plate) by photons, like coconuts being knocked off a fairground shelf by balls, was verified by the appearance of a photographic plate exposed to a very weak beam of light. The plate showed a patchwork of black spots where each photon had knocked out an electron, rather than a uniform gray exposure as would be expected if the light were a series of continuous waves.

Let me present how the photoelectric effect was understood before it was discovered that light also came in chunks of quanta. Then we can see what Einstein discovered and how this revealed that light was a particle and not a wave.

➤ When you shine ordinary white light on a plate, there is no current. In other words, no electrons are ejected no matter how bright the light.

➤ If you increase the frequency of the light to a *threshold frequency,* usually in the ultraviolet, electrons are ejected and current flows.

➤ When you go above the threshold, the higher the frequency of light, the electrons become more energetic.

➤ Increasing the brightness of the light does not increase the energy of the electrons. However, it does increase the size of the current.

Universal Constants

The **photoelectric effect** was first discovered by Heinrich Hertz in 1887. It works something like this. When you connect a battery to two zinc plates enclosed in a vacuum, you can send an electric current through the empty space between the two plates without the use of a wire, simple by shining light on one of the plates. One of the plates is connected to the negative terminal of the battery and the other plate is connected to the positive terminal. If you shine a beam of light only on the negative plate, a meter hooked to the circuit will show that an electrical current is flowing.

Universal Constants

The **frequency threshold** refers to metal plates involved in the photoelectric effect. Every metal responds to a certain frequency of light at which it releases electrons. If the frequency of light is below the metal's threshold, it won't release electrons and current won't flow. But as soon as the frequency level is high enough, electrons will be released and current will flow in the photoelectric circuit.

The **amplitude** of a light wave, or any form of electromagnetic wave, is simply the height of the wave. The amplitude along with the wavelength and frequency will define the characteristics of any wave.

Are you a little confused? Perhaps rightfully so. But let's go on, I think it will be cleared up soon. For one thing, what we call brightness of a light source is an informal word for its intensity. As you already know, if you increase the intensity of a light source, you increase the energy output. According to Maxwell, the greater the energy in a light wave, the greater the *amplitude*. So you can think of increasing the intensity as increasing the amplitude of the wave.

But this makes the whole photoelectric effect very mysterious. No matter how much you increase the energy of white light (by increasing the amplitude of the waves), you can't knock any electrons out of the zinc plate. However, if you do increase the frequency, which classically doesn't change the energy of the light, you do knock electrons out. And if you continue to increase the light's frequency, you don't increase the number of electrons knocked out, but you do increase the energy of the electrons. On the other hand, increasing the brightness, which classically does increase the energy of the light, doesn't increase the energy of the electrons, but it does increase the number of ejected electrons. Is this any clearer? Maybe once you see what Einstein did, it'll explain the difference between the classical position and the new quantum.

Einstein explained that by taking the equation at the heart of Planck's description of black body radiation, $E = hf$, and applying it to electromagnetic radiation, the photoelectric effect could be understood if light itself came in definite chunks, or quanta, each with an energy, hf. It takes one light quantum to knock one electron out of the metal and, for a particular frequency, all the light quanta have the same energy, so all of the liberated electrons have the same energy. In a brighter light of the same color (frequency), there are more light quanta, but each quantum still carries the same energy; so more electrons, still with the same energy, are liberated. And when the frequency is increased, for example from yellow to violet, the frequency is higher, so each light quantum carries more energy, and the liberated electrons move faster, even if only a few electrons are released. Got it?

Cosmonotes

Max Planck, Albert Einstein, and Neils Bohr made the first fundamental contributions to a new understanding of nature. Today the combined work of these three men, culminating in the Bohr model of the atom in 1913, is known as the old quantum theory. Three other individuals will later construct the new quantum theory, but that'll be found in another sidebar down the road.

After that lengthy discussion of the photoelectric effect, I hope you got the gist of the outcome. It boils down to this. The classical understanding of light as a wave couldn't explain why the predictions of how light should behave didn't match the results. Einstein's explanation of light behaving as a particle of energy (light quanta or photon), instead of a wave, produced results that matched predictions. The main problem with this was that it led to the wave/particle duality of light that everyone had a tough time accepting. But in any case, quantum mechanics was starting to take off, and Neils Bohr is our next stop.

Bohr's Atomic Theory

Neils Bohr is considered the grandfather of quantum mechanics. If Planck is the father of quantum mechanics, then Bohr must be older than him, which in fact he's not, they're not even related—another strange quantum anomaly.

We're now going to cover three of Bohr's most significant contributions (not all in this chapter, but the next chapter as well).

➤ Bohr's theory of the atom and its quantum structure

➤ The theory of complementarity, which helps explain wave/particle duality

➤ The Copenhagen interpretation, which is the standard explanation of what goes on in the quantum world

Having studied the ideas presented by Planck and Einstein, Bohr wanted to theorize about the quantum property of all forms of energy. In order to do that, he would have to explain how energy was released at the atomic level. He went about this by developing a better picture of the atom's structure. The current one that had been developed by Ernst Rutherford needed some tweaking to explain how atoms could emit light and yet not collapse in on themselves. As I hope you remember, light is created when energy is released from matter in the form of electromagnetic radiation. No one really knew how energy was released, they just knew that it was.

In the early 1920s, Bohr came up with a way to understand the stability and exactness of atoms using the analogy of standing waves. You can create your own standing waves by using a jump rope secured at both ends. If you pump energy into it and get it swinging, it can vibrate only in a certain number of predetermined ways. A violin string is another example. It can vibrate in its fundamental frequency, or twice, three times, or four times that frequency—in other words, its characteristic harmonics. It can't vibrate at two and one-half times that frequency. And if you can imagine an electron acting as a wave (remember that electrons can act as either a particle or a wave) within an atom in much the same way, you can see how it would be forced to assume only a certain number of predetermined vibrational states.

Let's look at another analogy. Remember our stair analogy discussed previously? Our child can't land safely or remain stable at step two and one-half, or three and one-third. She needs a minimum amount of energy before she can attain the next step or state. If she doesn't have quite enough energy to make it to step four, she'll remain at step

Mindwarps

Did you know that about 1,000 billion (10^{12}) photons of sunlight fall on a pinhead each second? Even when you look at a faint star, your eye receives a few hundred photons from that star each second. And some of them have traveled thousands of light years to get to you.

three. Atoms, too, act in the same way and will not absorb radiation unless the energy they receive contains the minimum to make the next quantum leap.

A child jumping down stairs also behaves somewhat like an electron changing states within an atom. In this case, she gives off energy to the floor as she jumps to a lower stair or state. But Bohr realized, like Planck and Einstein, that this energy can only come in chunks or quanta. A jump from step four to step two gives off two steps' worth of energy or a jump from step five to step two gives off three steps of energy.

An electron jumping to a lower state gives off its energy in the form of light. A jump from orbit or step five to orbit three might radiate quanta of red light; a jump from six to two might radiate more energy (higher frequency), so it could be a blue light. In terms of other types of electromagnetic radiation, a jump from orbit one to the ground might give off low energy radio waves, while a jump from orbit eight to the ground might give off high energy x-rays. This then was the way that Bohr described the quantum leap and how light was emitted.

But while this theory explained the quantum nature of atoms, some other unusual questions were raised, questions that have now become an accepted characteristic of quantum reality. For example, an electron can't exist between quantum states, not even for an instant. There is no such thing as in between. It's like jumping from one hour to the next without passing through the minutes in between, or disappearing from one end of a room to miraculously reappear at the other end. This quantum leaping in and out of existence can be very unnerving. How does a quantum state mysteriously materialize out of nowhere? How do you get from one place to another without crossing the territory in between?

Cosmonotes

We've used our imagination and have done some thought experiments to understand some of the theories we've discussed so far. And when it comes to quantum mechanics, we'll also be explaining these theories and concepts through familiar analogies, metaphors, and other images. In most cases, these will suffice to get the ideas involved across even though the theories are much more complex. However, this is the place where many of the physicists began to formulate their understanding of quantum interactions as well. When you're not really sure how something new operates, its often the best way to begin. Remember what Einstein said, "Imagination is more important than knowledge."

Mindwarps

The quantum nature of the universe is not limited to the subatomic world. It seems that some things have to come in whole chunks: children, snowflakes, memories, experiences, paintings, and a whole host of other things. This also seems to embody an irreducible yes/no, on/off quality that lies at the core of computer technology. Cultures, perceptions, beliefs, and even phases of life can often seem discretely separated as individual quantum states, which is why we can feel transformed when we move from one to another.

This unanswerable characteristic is true of the entire microcosmic quantum world. Virtually everything in the subatomic world is quantized. Not only energy and light, but also matter, momentum, electric charge, and many other exotic qualities of subatomic things, such as "strangeness" and "charm" (we'll get to these last two terms when we discuss properties of subatomic particles). An atom has to absorb energy by swallowing it whole and spits it back out in quantum chunks. This means the very stuff of the universe can't be smoothed out past a certain point, it has a grainy, lumpy texture.

So with this first of many unexplainable qualities under your quantum belt, I'll close this chapter with a quote from Richard Feynman, one of the most brilliant and gifted physicists/teachers of quantum mechanics. He said:

> There was a time when the newspapers said that only twelve men understood the theory of relativity. I do not believe there ever was such a time. There might have been a time when only one man did because he was the only guy that caught on, before he wrote his paper. But after people read the paper a lot of people understood the theory of relativity in one way or another, certainly more than twelve. On the other hand I think I can safely say that nobody understands quantum mechanics.

The Least You Need to Know

➤ The two cornerstones of physics today are quantum mechanics and general relativity.

➤ Max Planck solved the problem of black body radiation by explaining that energy is emitted in chunks of energy called quanta.

➤ Planck's constant is the ratio of a particle's energy to its frequency and is represented by the letter *h*—the formula that expresses this ratio is $h = E \div f$.

➤ By explaining the photoelectric effect, Einstein demonstrated that light is made up of packets of quanta that were later named photons.

➤ Neils Bohr's theories accounted for the stability of the atom, as well as how atoms emitted light.

Chapter 15

Chunks of Uncertainty

In This Chapter

➤ The dual nature of all matter

➤ The probability of quantum interactions

➤ It's all about measurement

➤ Complementary ideas

➤ Collapsing waves and cats

All quantum physicists will agree that they can't explain quantum theory and that it doesn't make sense. So why would they use something to define the fundamental structure of the universe if they don't understand what it is? Well, physicists who work in the field use the straightforward formulas handed down to them by the brilliant minds that developed them, but often don't understand why they work or even what they mean. But what is known about it has produced remarkably accurate results. And that's why it's used. Some even feel that quantum mechanics offers some of the most accurate numerical predictions that science has ever developed. And until something else comes along, the mathematical system used to define the quantum world is all that is available. For now, let's pick up where we left off in the last chapter and follow the unfolding of quantum cosmology.

Waves of Matter

Neils Bohr came up with a description of how light was radiated from inside an atom. He also has shown why an atom was stable. His explanation of what the structure of the atom was like was close to but not quite like the one envisioned by Rutherford that we touched upon earlier. Remember that he theorized that the atom's structure was very much like a planetary system with the electrons orbiting the nucleus the way planets orbit the Sun. Bohr adopted this basic configuration, but couldn't imagine the electrons orbiting the nucleus as some cosmic cloud that was indefinable. So he had arranged the orbiting electrons into layers or shells. Have you seen Russian nesting dolls or Chinese stacking boxes? In those you have one complete doll or box contained within another. Every time you open up one there's a smaller one inside. The shells Bohr described were just like that, except that each shell had a specific number of electrons.

Bohr was able to calculate mathematically the diameter of each electron orbit along with the maximum number of electrons in each shell. The *angular momentum* of the electron in orbit was counteracted by the attraction of the nucleus. In other words, since unlike electrical charges attract each other, the positive charge inside the nucleus attracted the negative charge of the electron. This theory explained the structure of the atom and why it remained stable.

A French Prince Discovers Matter Waves

In 1923, a graduate student at the Sorbonne in Paris, Prince Louis de Broglie (1892–1987), introduced the remarkable idea that particles may exhibit wave properties. He had been strongly influenced by Einstein's arguments that light had a dual nature. He was also deeply impressed by Einstein's particles of light that could cause the photoelectric effect (knock electrons out of metal) while also producing the interference patterns caused by waves as in the double slit experiment. He proposed one of the great unifying principles in quantum physics. He was convinced that the wave/particle duality discovered by Einstein in his theory of light quanta was a general principle that extended to all forms of matter. In other words, the propagation of a wave is associated with the motion of a particle of any kind—photon, electron, proton, or any other.

De Broglie wished to develop a mechanical explanation for the wave/particle duality of light and to extend this to all forms of matter. He needed to find a mechanical reason for the photons in the wave to have an energy that was determined by the frequency of that wave.

De Broglie noticed a connection between the angular momentum of the electron in a Bohr orbit and the number of *nodes* in a standing wave pattern (remember you created those with the jump rope in the last chapter). The orbiting electrons could only have one unit of *h* (Planck's constant) or two units, etc. Could these discontinuous changes in the electron's angular momentum, these changes in the amount of *h* allowed, be due somehow to a similar change in standing wave patterns?

Universal Constants

Angular momentum can be thought of as momentum moving in a circle. It's what keeps a ball moving on the end of a rope as you spin it over your head. In classical physics it is defined by the mass and the speed at which it is spinning. In quantum mechanics, the angular momentum, like everything else, is quantized.

Cosmonotes

One of the reasons de Broglie wanted to develop a mechanical model was to show that the theories developed so far could be verified by experimental means. Up until then, the Rutherford–Bohr atomic model was just theory; no one really knew whether the atoms looked like that. If he could show that his predictions could be confirmed by experiment, it would solidify the new ground on which quantum theory was developing.

De Broglie realized that the Bohr orbit could be seen as a circular violin string, like a snake swallowing its own tail. Would the orbit size predicted by his standing matter waves correspond to Bohr's calculated electron shells? What would his wave do if confined to a circle? Well, what he discovered was that his matter waves fit Bohr's orbits exactly. And when he calculated the wavelength of the lowest orbit, he discovered another astonishing mathematical connection between the wave and the particle. The momentum of the orbiting electron equaled Planck's constant divided by the wavelength.

Universal Constants

When you create a standing wave with a jump rope, each point on the rope where there is no movement, a resting place, is called a **node.** It's the point at the end of each wave and also the point between the waves that are moving up and down. For example, there are two nodes on the lowest frequency standing wave (a half wave), the two endpoints of the rope. The next higher frequency, (a whole wave) has three nodes, the ones on each end and a third in the middle, which is the point that separates the crest from the trough. The next higher frequency ($1^1/_2$ waves) has four nodes, the two end points and two in the middle and so on. Get the idea? As the number of nodes on the rope increases, the frequency of the standing wave increases. If this were a vibrating violin string, the pitch would also increase.

Just a Little Math Won't Hurt

De Broglie had discovered a new formula, one as important and startling as Planck's formula. It stated that the momentum of a particle p was equal to Planck's constant h, divided by the wavelength λ. That is, p = h ÷ λ. I would like to walk you through the math for this formula because it was a brilliant insight. It's just a matter of substituting some letters and rewriting Einstein's famous, $E = mc^2$.

We begin with

$E = mc^2$, which is the same as E = mc × c

Because mc is mass times speed, or in other words momentum (p), we can substitute p for mc. So,

E = p × c

Because the term c stands for speed of light, we can substitute it with f (frequency) times λ (the symbol lambda, which represents wavelength) because that's the formula for finding the speed at which waves travel. Then,

E = p × f λ

This is almost the same formula just different terms.

Now if we use the Planck/Einstein equation that relates a particle's energy to its frequency, or E = h × f (you were introduced to it as h = E ÷ f to show you how Planck's constant was derived, but we just rewrote it in another form) we get the following:

h × f = p × f λ

We do a little simple algebra (divide both sides by f λ, which cancels out the two f λ's on the right and the two f's on the left):

h × f ÷ f λ = p × f λ ÷ f λ

And we get h ÷ λ = p or turned around p = h ÷ λ.

That means that a particle (p) is equal to Planck's constant (h) divided by the wavelength (λ). Cool, huh? Well maybe not to some of you. But don't worry if you didn't follow it, you'll still be able to understand why it was important. With this new mathematical discovery, Bohr's orbits could be explained. Each orbit was a standing wave pattern. The atom could be understood as a finely tuned instrument (maybe Pythagoras's and Plato's music of the spheres wasn't that far off). These mathematical relations balanced the tiny electron into a tuned standing wave pattern. Orbits had determined and fixed sizes in order that these distinct, quantized wave patterns could exist.

Black Holes

The traditional establishment in any area of study, whether it's medicine, physics, or even art and music, often takes a dim view of new ideas or forms of expression. More than one scientist's theories have been scoffed at and ridiculed because they went beyond the accepted understanding of how things are supposed to be. Einstein met with disbelief at first, as did Louis de Broglie. Many considered his explanation of matter waves as crazy and absurd. Of course, he was later proved to be correct when experiments were done that showed that matter did indeed have a wave property. It's a good thing that the innovators and visionaries of society don't pay attention to conventional ideas, otherwise we could still be making tools from stone.

I hope that these last few pages have given you a good idea of de Broglie's contribution to quantum mechanics. However, if you're still a little lost, the main thing you need to know is that all matter, everything that exists, has as its core characteristic the wave/particle duality. This dual nature of matter is probably the one significant thing that most quantum cosmology rests upon.

Mindwarps

Erwin Schrödinger had a disdain for convention, especially regarding sex, but it extended beyond that. Many times when he attended a conference, he would walk from the train station to the hotel where the delegates stayed, carrying only a rucksack and looking like a tramp. It always required a great deal of argument at the reception desk before he could claim his room. But this same disdain for convention made him very popular with his students and he was highly regarded as a teacher.

More Wave Mechanics

While de Broglie's theories offered a picture of what was going on inside of an atom, more was needed to explain the shifting patterns of the wave when it changed its energy and emitted light. Erwin Schrödinger, an Austrian physicist, found a mathematical equation that explain-ed the changing wave patterns inside an atom. Don't worry—I'm not going to take you through the math this time. As a matter of fact, I just want to give you a brief overview of his wave mechanics because I want to spend more time on the really weird stuff that's just around the corner.

Schrödinger, like de Broglie, used the analogy of a vibrating violin string to explain his mathematical equation. The movement of an electron from one orbit to another lower energy orbit was a simple change in notes on a violin. As the violin string undergoes such a change, there is a moment when both harmonics (sound produced by the notes) can be heard. This results in the well-known experience of harmony, or as wave scientists call it, the phenomenon of *beats*. The beats between two notes are what we hear as harmony and this really produces a third sound. The vibrational pattern of the beats is determined by the difference in the frequencies of the two harmonics.

This was just what was needed to explain the observed frequency of the light waves of photons emitted when an electron in the atom underwent a change from one orbit to the other. The light was a beat, a harmony, between the upper and lower harmonics of Schrödinger's and de Broglie's waves. When we see light emitted in this process, we are observing an atom singing in harmony.

Universal Constants

A **beat** has a very different meaning than what we're accustomed to when it's talked about in wave mechanics. In this case, it refers to combining two waves of different frequencies that produces an additional frequency equal to the difference between the two. You can think if it as another word for the auditory experience of harmony.

So what's the big difference between de Broglie's and Schrödinger's waves? It essentially boils down to the following:

➤ De Broglie's waves only accounted for the electron as a one-dimensional wave. Schrödinger's wave *function* described the waves in three dimensions.

➤ De Broglie's explanation was a mechanical one. Schrödinger's was purely mathematical.

➤ De Broglie's theory utilized the quantum nature described by Einstein, while Schrödinger disliked the idea of discontinuous quantum jumps. Through his mathematical explanation he tried to return to a more classical understanding.

And after all was said and done, Schrödinger's equations became the preferred method for solving quantum interactions. And it opened the door for another cornerstone of quantum mechanics, probability theory.

Born of Probability

Schrödingers's picture of the atom relied on a complicated and hard to imagine wave function. Nevertheless, it worked, and the atom's electron was indeed a wave. The atom radiated not because its electrons leaped from orbit to orbit, but because of a continuous process of harmonic beats. Light was emitted when the atom "played" both the upper and lower energy frequencies at the same time. The difference between the two electron matter/wave frequencies, which also corresponded to Bohr's conception of the atom as the difference in the electron's orbital energies, was exactly the frequency of the light observed.

Eventually, Schrödinger's musical picture of the musical atom was replaced by a new one, but his mathematical equations remained intact. The problem with a pure wave model was that it didn't account for the electron's other feature, that of a particle. No matter how the wave shook and danced, there had to be a particle somewhere. Max Born (1882–1970), another German physicist, was the first to provide an interpretation of this particle discontinuity. He realized that the wave was not the electron. It was a wave of *probability*.

We can now add the notion of probability as another chief feature of quantum mechanics to de Broglie's contribution that all matter exhibits wave/particle duality. Probability is a strange characteristic to have as a fundamental building block of the universe. It somehow seems to reflect an incomplete picture of our understanding of how the quantum world operates. This is exactly how Einstein felt and was reflected in his famous statement, "God does not play dice with the universe." He never fully accepted this flaw of quantum mechanics and always felt a more complete picture would be developed down the road. However, after more than fifty years of experiments, it seems that matter must be described in an essentially probabilistic manner, at least at this point in time.

Cosmonotes

In the last chapter and to this point, we've covered the quantum ideas and theories of some of the key scientists. They've been presented in a way so that you could see how each one, more or less, unfolded into or gave birth to the next. Before going further, it's a good idea that you to have a clear idea of how this stream of development occurred. Because from here on in the rest of the theories will get a little more unusual. So it's a good idea to have a clear picture of the state of quantum mechanics right now.

The difficulty with an electromagnetic wave of any kind is that it's impossible to locate the position of any one particle, like an electron, within it. Born realized this and developed a way to determine its probable location. By calculating the magnitude or intensity of a wave, he found that in places where the wave was large there was a greater probability for the electron to be found. And where the intensity was small, the probability was less.

Universal Constants

Probability is something we're all familiar with from tossing coins at sports events to betting in casinos. It is also the basis for statistical analysis and when used in that framework has a very precise mathematical system. It can be usefully applied to predict certain outcomes, but also has the ability to be manipulated to reflect whatever outcome someone wants. It all depends on what you want to use it for. As a tool in quantum mechanics, it is the only method that can show us enough about the microcosmic universe to understand it at this point.

Probability Leads to Uncertainty

The role of probability became a crucial part of quantum mechanics. And because it did, it became the fount from which sprang other core theories and concepts that defined the quantum world in strange and exotic ways. The deeper physicists probed into the quantum world, the more the explanations veered from normal everyday experience. It seemed as though the fundamental structure of the universe operated in ways that were mysterious and profound at the same time. In the upcoming pages remaining in this chapter, we'll look at some of these ideas so you can see for yourself just how weird it can all get.

The only way a scientist can start to understand something is to describe it, measure it, and name it. You can't begin to measure something until you make some assumptions about what that something is. You can't measure the amount of beauty in a contestant without defining in some sense what beauty is any more than you can measure success, distance, motion, time, or intelligence without defining what it is. And measurement itself is as much a mode of perception as seeing and hearing. Reality and the way we measure it begins and ends with ourselves. And as Werner Heisenberg said, "We have to remember that what we observe is not nature herself, but nature exposed to our method of questioning."

Cosmonotes

Richard Feynman believed that the double slit experiment encapsulated the "central mystery" of quantum mechanics. He said it is "a phenomenon which is impossible, absolutely impossible, to explain in any classical way, and which has in it the heart of quantum mechanics. In reality, it contains the only mystery ... the basic peculiarities of all quantum behavior."

This may seem an unusual idea, because after all what is more scientific than measurement? Yet as we'll soon see, our ability to measure what goes on in the quantum world is what causes all of these theories to be developed in the first place. Let's begin with a return to the double slit experiment.

Back in Chapter 10, "The Dual Nature of Light," we saw how the act of measuring a beam of light defined it either as a wave or particle. We could not measure it as both a wave and a particle at the same time. If we extend that idea to include two other measurements—position and momentum—we'll have discovered Heisenberg's uncertainty principle, another cornerstone of quantum mechanics. Let me explain. Position is a property of a particle while momentum is a property of a wave. Since we can only know if a beam of light or any form of matter is either a particle or a wave, we can therefore only measure its position or its momentum, but not both.

To sum it up, the Heisenberg uncertainty principle can be expressed as follows:

➤ You can't accurately measure the position of a subatomic particle unless you're willing to be uncertain about its momentum.

➤ You can't accurately measure the momentum of a subatomic particle unless you're willing to be uncertain about its position.

This summary also shows that it all comes back to the person doing the observing. Not only does this principle reflect the uncertainty of what you can measure, but also that the very act of measuring affects the outcome of the experiment. It is not possible to separate the experiment from the experimenter. To put it succinctly, "To observe is to disturb."

They're Complementary After All

Heisenberg's uncertainty principle led Neils Bohr to father the concept of complementarity. Bohr said that the reality of particles required *complementary* descriptions— more than one point of view. It doesn't matter that you can't measure both motion and position at the same time; you can't see both sides of a coin at the same time either. As long as we insist on looking at the subatomic world with our everyday perspective, (and what other choice do we have?), we will be stuck with looking at nature one dimension at a time. Take a look at the following figure to see exactly how the principle of complimentarity operates.

Can you see both the chalice and the faces at the same time?

When you look at the image, you can either see the outline of the chalice or the outline of the two faces, but you can't see both at the same time. This illustrates the fundamental paradox of the wave/particle duality of matter and shows you the basis for Heisenberg's uncertainty principle and Bohr's concept of complementarity.

190

Universal Constants

Complementary ideas are opposing ideas that add up to much more than the sum of their parts. They complement each other like night and day, male and female, yin and yang. One helps to define the other and each is equally important as the other. Waves and particles are complementary ways of describing the nature of light as well as all energy and matter.

For centuries, people have argued over whether light was a wave or a particle (remember Newton's and Huygen's opposing views in Chapter 8, "Physics 101"?). In today's world, we know that light is both and to argue about it is like asking whether the color of the sky is blue or whether it has mathematical properties. Each is true in the proper context. It is rather like the sides of a box or the facets of solving a problem. What you see depends on what you're looking for, which is why light and all energy and matter can show up as quantum chunks in some experiments and as waves in others.

Complementarity makes it easier to accept the innate limits on perception and measurement. Each way of seeing only goes so far. In the same way that we need two eyes to see depth (the combination of two distinct images), we need more than one perspective to see something in all its dimensions.

Wonderful, Wonderful Copenhagen

We now have a number of concepts fundamental to the inner workings of the quantum universe. All of these come together in Bohr's *Copenhagen interpretation*. This is the standard interpretation of what goes on in the quantum world. It held sway from the 1930s to the 1980s and is still taught in most textbooks and many university courses. It is by no means the only way to interpret quantum mechanics, but it is the most popular, owing much of this to the forceful personality of Neils Bohr.

Here's the basic idea of how it works. Let's take an empty box. If we put an electron in the box, even if we don't where it is, it has a definite location. The Copenhagen interpretation says that the electron exists as a wave filling the box and could be anywhere inside. At the moment we look for the electron, the wave function collapses at a certain location. This is similar to an electron in the double-slit experiment. As soon as you observe it, the electron stops behaving like a wave and behaves as a particle.

Universal Constants

The **Copenhagen interpretation** got its name from the city where Bohr worked. It is one of the central theories of quantum mechanics. It basically states that there is no meaning to the existence of a quantum particle unless it is observed. In other words, until you see it, it doesn't exist. The Copenhagen interpretation stresses the role of experimentation to understand the quantum world. Bohr insisted that the only thing we can know for sure is what we can measure with our instruments.

Now if we slide a partition in the box without looking, the electron must be in one half of the box or the other. The Copenhagen interpretation says that as long as we don't look, the electron wave still occupies both halves of the box and only collapses on one side of the barrier when we look inside. As long as we don't look, even if we move the two halves of the box far apart, the wave still fills both boxes. Even if the boxes are moved light years apart, it is only when we look into either one that the electron wave collapses, instantaneously, and the electron "decides" which box it is in.

The idea that the electron is in both halves of the box at the same time is based on probability. And according to Heisenberg's uncertainty principle (by observing it we change the outcome) leads to the paradox that somehow the electron knows which box to be in, even if light years apart. How can this information be passed instantaneously when nothing can travel faster than light? It's not called a mystery for nothing.

Some physicists thought this idea was rather absurd. One in particular, our friend Schrödinger, put together his famous thought experiment to show just how nonsensical this whole interpretation was. It's known as "Schrödinger's Cat." Let's take a box and inside we'll place a radioactive source, a Geiger counter, a hammer, and a glass vial filled with poisonous gas. And, of course, we need a live cat, too (don't worry—no real cat was ever used in this experiment). After a period of time, radioactive decay takes place, the Geiger counter measures this and triggers a device that trips the hammer, which then breaks the vial. Poison is released and the cat dies.

Let's say that in this particular instance, quantum theory predicts that there is a 50 percent probability of one decay particle for each hour from the radioactive source. After one hour, the decay has triggered the entire apparatus and the cat is now either alive or dead. Within the framework of the Copenhagen interpretation, exactly one hour after the experiment began, the box contains a cat that is both alive and dead—a mixture of two states. This is just like the two states of a quantum particle, either it's a wave or a particle. It only becomes a particle after it is observed. Taking this analogy to the cat, it isn't dead until the box is opened and the cat is observed. Until then it's still alive and dead at the same time.

Mindwarps

The idea of the passage of information instantaneously is called action at a distance. It's the same as Newton's idea of how gravity operated—remember? Einstein hated the whole idea of this "spooky" action at a distance. It went against the fact that light was the top speed of anything in the universe. How could two waves light years apart know which box was being opened so that the correct box's wave could collapse into a particle? It's why Bohr theorized that the existence of a quantum particle has no meaning until it is observed.

You can see how strange this is. How can the cat be both alive and dead at the same time? Schrödinger and others felt this paradoxical state showed the absurdity of the Copenhagen interpretation. But regardless of how strange it seems, it's still one of the basic ways used to describe the quantum universe. And with Schrödinger's cat either alive or dead, we'll close this introduction to some of the unusual characteristics of quantum interactions. We're not completely done with quantum mechanics because there are still some aspects of it that we need to look at when we move into string theory. But for now, I think you have a good idea of some of the more significant and strange contributions that developed into the world of the quantum. In the next chapter, we'll take a bird's eye view of the entities that inhabit the microcosm and extend our understanding of quantum cosmology.

The Least You Need to Know

➤ All matter has the wave/particle duality as a core characteristic.

➤ Heisenberg's uncertainty principle states you can know the position or the momentum of a quantum particle, but not both at the same time.

➤ In the quantum world, you can't separate the experiment from the experimenter: "To observe is to disturb."

➤ Waves and particles are complementary ways of describing the nature of light, energy, and matter.

➤ Probability, uncertainty, complementarity, and paradox are all chief features of the quantum universe.

Chapter 16

Forces, Particles, and Some Cosmological Glue

> **In This Chapter**
>
> ➤ The world of the atom
>
> ➤ Subatomic interactions
>
> ➤ The four forces of the universe
>
> ➤ The eightfold way
>
> ➤ The standard model of particle physics

We've spent the last two chapters examining the weird and wonderful world of quantum interactions. I think you've learned some surprising things about how different the quantum universe operates from our everyday experience of the world. But while we've looked at how particles and waves behave, we haven't taken a close look at how many different types of particles and forces actually make up the microcosm of the quantum playing field. By understanding the microcosm, you will learn how the physicists began to develop theories about the macrocosm (our universe) including such theories as the big bang. In our study of cosmology, we've reached a point where the fundamental structure of the universe will be described by understanding the fundamental structures contained in very small particles.

We're now going to spend a chapter discovering the world of particle physics. From atoms to quarks and fermions to bosons, we'll delve into the subatomic universe and stare these microcosmic inhabitants in the face. Linear accelerators and cyclotrons will split particles into smaller and smaller pieces, with the hope of some day finding the tiniest form of matter. By the time we're through, you'll know that strange and charming are terms that can apply to particles as well as people. So strap yourself to the nearest super collider and let's go find the standard model of this microscopic cosmos.

An Atomic Review

Do you know the first particle that was discovered? We've talked about it a few times already … no, not the atom—that's too big. It was the electron. However, I don't want to spend a lot of time talking about the atom and its parts—there are too many pieces in this puzzle to put together—but that's a great place to start. The key to the structure of the atom, what it's composed of, why the nucleus stays together, and the mass of each of its parts are all very important. It's where particle physics began. So let's just spend a little time reviewing what we know about the atom and take it from there; everything after the atom just gets smaller and smaller.

In 1897, J.J. Thompson, a young English physicist, had been performing a number of experiments with *cathode* rays, trying to find out if they were really particles. He built a simple apparatus in which the cathode rays were directed across a region between two electrically charged plates, and in this region there was also a magnetic field. The important result that Thompson obtained was that the cathode rays were particles and that they were attracted toward a positively charged plate. This meant that the particles carried a negative electrical charge (and I know you know that opposite charges attract, I just wanted to remind you).

Universal Constants

Before scientists knew about particles, most believed that electromagnetic radiation was some type of rays—such as x-rays, cathode rays, etc. The term **cathode** refers to the negatively charged plate inside of a cathode ray tube, or CRT. The positively charged plate is called the anode. Electrons move from the cathode to the anode because they have a negative electrical charge. CRTs are the basic apparatus behind TVs, radar screens, computer screens, and oscilloscopes.

This particle that Thompson discovered is now known as the electron. It has a mass of 9.1×10^{-28} gram. That's extremely small. And once it was properly identified and labeled, it was realized that the electron was a very important particle. Every electrical current, whether it's a man-made circuit or a nerve in your body, is simply a flow of electrons. And where else could the electron come from but the interior of what was once thought of as the indivisible atom? The existence of a negatively charged particle that could be taken from the atom implied that there must also be a positively charged segment left behind, and this in turn implied that the atom must have structure. If this was so, there must be a type of matter more fundamental than the atom. The electron was the first example of matter from the subatomic realms.

All this may seem a little after the fact since we've already talked about electrons and the structure of the atom. But for our purposes, it's good to start at the beginning so that we can trace the chain of events and ideas that developed into the science of particle physics.

The Nucleus and the Proton

In the early 1900s, the ideas current at the time envisioned the atom as a large, diffuse, positively charged chunk of material in which the electrons were embedded like raisins in a bun. Thanks to Ernest Rutherford, whom you met in Chapter 10, "The Dual Nature of Light," the atom took on the familiar shape of a planetary system that was later adopted by Neils Bohr. Once the existence of the nucleus was established, scientists began to question its composition. After a few more experiments in which Rutherford observed the emission of certain particles from nuclear collisions between helium and hydrogen nuclei, he discovered and named the new particle the proton, which in Greek means "the first one."

The proton is a particle that has a positive electrical charge. The magnitude of the charge is precisely equal to the magnitude of the negative charge of the electron. However its mass is about 1,836 times that of the electron. An interesting point about protons is that the number of protons in the nucleus of an atom also reflects its atomic number, or its numerical order, in the periodic table of elements. One other insight that Rutherford had was that there was probably another constituent to the heavier nuclei. In other words, he thought there was an electrically neutral particle as massive as the proton in the nucleus as well. He arrived at this conclusion by noting that most atoms apparently weighed about twice as much as you would expect them to if you added up the masses of the protons and electrons. He called this hypothetical particle the neutron, which was eventually discovered in 1932.

Size in the Microcosm

I'd like to put the subatomic world in perspective for you. Molecules are huge compared to atoms and the particles that make them up. And using normal numbering would take too long to write their values out. For example, a typical nuclear diameter may be 0.000000000000006 meters. There

Mindwarps

Not only scientists, but also writers of science fiction adopted the analogy between the nuclear atom and the solar system. This idea served as a plot for innumerable movies, comics, and stories during the 1930s and '40s. And in that sense it has also become the accepted folklore of modern culture, even though in reality it doesn't look like that at all.

197

had to be an easier way to write this. That's where something called scientific nota-tion comes in. This is a method that was developed to write large numbers in powers of ten. So the number above would be written as 6×10^{-14} meters in scientific notation. Using this method, let's look at the relative size of these three subatomic particles.

➤ The electron is the negatively charged particle that "orbits" the nucleus. It has a mass of 9.1×10^{-28} grams.

➤ The proton is the positively charged particle in the nucleus of the atom. It has a mass of 1.67×10^{-24} grams. It has a mass 1,836 times greater than the electron and its radius is 8×10^{-16} meters.

➤ The neutron is the neutral particle in the nucleus of the atom. It has a mass al-most identical to that of the proton. And it is 1,839 times as heavy as the elec-tron.

➤ The radius of a typical nucleus is 3×10^{-15} meters, while the radius of a typical atom is 3×10^{-10} meters. That means that the diameter of the nucleus is about one hundred thousandth of the diameter of the whole atom.

➤ The volume of the nucleus, the space it takes up, is one trillionth of the whole atom.

To relate this to a more understandable perspective, let's increase the size of the nu-cleus to about a foot, maybe the size of a bowling ball or medium-sized melon. How big would the atom be? Close to twenty miles in diameter. So if the nucleus of our atom were as large as a bowling ball, the rest of the atom would consist of ten pea-sized electrons scattered around a sphere twenty miles across with the bowling ball at the center. Imagine putting a bowling ball in the center of a city and then scattering ten peas throughout the rest of the city and you'll have some idea of how empty an atom really is.

The Fundamental Forces of Nature

As you can see, most of the matter in the atom is contained in the nucleus. We also know that the nucleus has a positive electrical charge. But wait a second, don't like charges repel each other? If the nucleus is all protons and neutrons (which as you know are neutral in charge), what prevents the protons from repelling each other and the nucleus from flying apart? Obviously there must be some kind of cosmological glue that keeps the whole thing together. The nature of this force is rather mysteri-ous. It has to be a force that can overcome the repulsive force of an electric charge.

The repulsive electrical force is so powerful that this force must be many orders of magnitude stronger to keep the nucleus together. It is actually more powerful than any force we deal with at the everyday level. Physicists have named this the *strong force*. And part of the development of particle physics has been an attempt to under-stand what the strong force is and how it is generated. You'll find out about this a lit-tle further into the chapter.

Mindwarps

To give you an idea of how strong the force is that holds a nucleus together, let's scale up the size of the proton. If we made it a foot in diameter, with the center of the protons about 18 inches apart, how large would this force be? Well, if we embedded these two oversized protons in the strongest metal alloy known, the electrical repulsion would be so strong it would tear this metal apart as if it were tissue paper. So this force has to be strong enough to overcome these powerful forces that seek to push apart.

Universal Constants

The **strong force** is one of four fundamental forces under which the universe operates. It is also known as the strong interaction; it operates within the nucleus of the atom, keeping the nucleus together and overcoming the tendency of the positive electric charge to blow it apart. The strong force is about 100 times stronger than electromagnetism (over the size of the nucleus), which is why there are roughly 100 protons in the largest stable nuclei.

The Neutron Decay

As mentioned in the previous section, the mass of the neutron is pretty close to that of the proton, just a little more. In fact the mass is very close to the combined masses of a proton and an electron. But the neutron exhibits a property not found in the proton or electron—instability. If we could put a neutron on a table and observe it, it wouldn't be there for very long; it would disappear. And in its place you would find a proton, an electron, and another type of particle, which I will get to shortly.

In the language of particle physics, a neutron outside of a nucleus, "decays," a term you've already been introduced to. This process of decay is what it means for a particle to be unstable. As we get further into looking at the whole slew of elementary particles you'll see that virtually all of them share this characteristic of instability. And if you remember back in Chapter 12, "The Relative Nature of Space and Time," where I discussed the carbon-14 dating method, we saw that this was exactly the form of decay that allowed scientists to date the age of any material.

Let's take a closer look at neutron decay. When the decay occurs, two charged particles are created—a proton and an electron. This means the total electrical charge after the final decay is 0, which is exactly the charge on the neutron. The total electrical charge of the process has been conserved, or in other words, remains the same as before the decay took place. However, upon very close measurement, physicists found that the amount of electrical charge left after the decay was slightly different than before. Well, they couldn't throw out the laws of conservation, so there had to be something else that could account for the missing amount of energy.

Cosmonotes

In every reaction involving elementary particles, the total electrical charge is the same before and after the reaction. This is called the conservation of electrical charge. Conservation laws play an extremely important role in physics. For example, there are laws that tell us that the energy of a system has to be the same before and after every reaction, which is the basis for Einstein's famous formula, $E = mc^2$, which in turn shows the equivalence of energy and mass. And there are other laws that tell us the same thing about momentum.

Neutrinos and the Weak Force

In 1934, Italian physicist Enrico Fermi (the man who built the first nuclear reactor at the University of Chicago) put together the first successful theory of decay that explained where the missing energy went. He theorized that there was another particle that must be electrically neutral (otherwise it would have been detected earlier) that he named the neutrino. If such a particle existed, it would carry away the missing energy of the neutron decay and the laws of conservation would be preserved.

Mindwarps

One way of gauging the neutrino's lack of interaction with other matter is to give you a good example. If you were to make a lead plate neither meters thick, nor kilometers thick, but light-years thick, you might stand a chance of getting the neutrino to interact with an atom in it. Even if you were to run a lead tube from the Earth to our nearest star, Alpha Centauri, and started a neutrino down the tube today, four years from now it would emerge without disturbing a single atom of lead in the tube.

In particle physics the only way for any particle to be detected is for it to interact with something and cause a change. Well, the neutrino posed certain difficulties for physicists because the neutrino, like a photon, has no mass. You can't detect an interaction when the particle has no mass to interact with anything. And the neutrino doesn't readily interact with other matter at all.

Eventually in 1956, almost 20 years after Fermi first theorized the neutrino's existence, they were finally detected in an experiment with a nuclear reactor. The neutrino is now a fully accepted particle routinely produced for experimental purposes in many large accelerators around the world. Reactions in which the neutrino is created through particle decay occur on a relatively slow time scale and are given the name *weak force* or *weak interaction*.

Let's pause for a moment and tally our sheet of forces and particles. So far we've looked at two of the four fundamental interactions of nature—the strong force and the weak force. We've also examined the electron, proton, neutron, and neutrino. In addition to these four particles, we should also add a fifth, the photon. As you know, the photon is a particle of light and therefore also a constituent of radiation. Almost all of the particles that you'll meet from here on will be unstable, only the electron and the proton can exist by themselves for indefinite periods of time. All other particles decay into some combination of photons, neutrinos, protons, and electrons. And just to clarify things, we have two forces and five particles so far. Before we move into all of the other families of particles and interactions, let's complete this section with a look at the last two of the four fundamental forces.

Universal Constants

The **weak force** or **weak interaction** occurs during a process known as beta decay. Beta decay is just another term used by physicists for the breakdown or decay of a neutron. This occurs when a neutron, either in the nucleus of an atom or as a free neutron, transforms through this weak force into an electron, a proton, and a neutrino (in actuality it's called an electron antineutrino, but let's not confuse the issue). It is the second of the four fundamental forces operating in our universe.

The Forces You Already Know

There are only four forces that are known to operate between elementary particles. Two of these we just covered, the strong force and the weak force. The other two, gravity and electromagnetism, are familiar to us because, not only have we covered them, but they also operate in the everyday world. Gravity is the weakest of the four forces and was the first to be discovered. But even though it is the weakest of the four, its range is infinite, while the strong and weak forces are limited in range to the nucleus of the atom. Electromagnetism is much stronger than gravity, but both electricity and magnetism come in two varieties—positive and negative charge, north and south poles. These varieties tend to cancel each other out, reducing their overall influence. But as you know, Maxwell combined these two separate forces into one: electromagnetism. This force holds atoms together (electrons in their probable positions around the nucleus) as well as molecules, but is overcome by the strong force that binds the nucleus together. This all might be explained a little better by the following chart. You can compare the four forces and their relative strengths.

The Four Forces

Name	Strength Compared to Strong Force	Effective Range in Centimeters	What It Does
Gravity	6×10^{-39}	Infinite	Holds planets, stars, and galaxies together
Weak	1×10^{-5}	1×10^{-15}	Causes particles to decay
Electromagnetism	7×10^{-3}	Infinite	Holds atoms together
Strong	$1 \times 10^{0} = 1$	1×10^{-13}	Holds atomic nucleus together

Cosmonotes

It's interesting that the fundamental forces of the universe happen to manifest as the number 4. If you remember back to the first chapter when I discussed the role of certain numbers, such as 4 and 7, which were used by ancient cultures to define the cosmological structure of their universe, the tribal Native American people believed that "the Great Spirit caused everything to be in fours." Maybe they already knew something that has taken science centuries to find out.

Cosmic Rays, Greek Particles, and Accelerators

In the last sections of this chapter, we'll be taking a look at the plethora of particles that go to make up what is known as the standard model of particle physics. Contained within this model are the particles responsible for carrying out the work of the four fundamental forces we've already discussed. Since Einstein's famous equation showed that matter and energy are equivalent, physicists describe particles in terms of their energy content rather than just their mass. This unit of measure is called the *electron volt* (eV). This is an extremely small amount of energy, less than the amount used by a bug to flap its wings.

Before accelerators were built to study the elementary particles that were detected in their subatomic collision experiments, nature had provided a natural laboratory in which the interaction of elementary particles could be studied. *Cosmic rays* are elementary particles that travel to earth from our sun and other stars in our galaxy. They are mostly protons and they interact with other particles in our atmosphere as they make their journey to the surface of the Earth. The collisions that are created by these particles cause a cascading effect. As one particle hits another, more particles are created and these in turn hit other particles and so on.

Seeing an elementary particle is not an easy thing to do, so an apparatus was built to observe these cosmic ray collisions. In the 1930s, the Wilson cloud chamber was invented to fill this need.

Universal Constants

Cosmic rays are energetic particles from space, including electrons and protons, some of which interact with the nuclei of atoms in the atmosphere of the Earth to produce showers of secondary particles. Before the development of particle accelerators, cosmic rays provided physicists with their only source of high-energy particles to study.

When a particle travels through the gas in the chamber, it leaves a trail of vapor, similar to the trail from jets in the sky. Out of these experiments came the discovery of the first particle of *antimatter,* the positive electron, or positron. The cloud chamber proved to be a very useful observation device, because more and more particles were discovered.

Universal Constants

An **electron volt (eV)** is a measure of energy introduced in 1912. It's equal to the energy gained by one electron when it is accelerated across an electric potential difference of 1 volt. (For you budding scientists out there, this would read $1eV = 1.602 \times 10^{-19}$ joules.) Because this unit is so small, it is more commonly encountered as keV (thousand electron volts), MeV (million electron volts) or CeV (billion electron volts). If you dropped this book from a height of about 3 inches, it would be accelerated by gravity to kinetic energy of about 1 billion electron volts. A 100-watt light bulb burns energy at the rate of 6.24×10^{20} per second. But it takes only 13.6eV to knock an electron right out of an atom of hydrogen and the energy of particles produced in radioactive decay are typically several MeV. This gives you an idea of the different energies associated with chemical and nuclear processes.

Universal Constants

Antimatter is a form of matter in which each particle has the opposite set of quantum properties (such as electric charge) to its counterpart in the everyday world. When an antiparticle meets its particle counterpart they annihilate each other, converting their mass into energy. There are antimatter counterparts for all matter of particles—antiprotons, antineutrons, antineutrinos, and so on.

Mu and Pi

Three of the more significant particles that were discovered with the use of a cloud chamber all were assigned Greek letters. (They were all members of a new classification of particles based on nationality ... not really, physicists just love to use the Greek alphabet when assigning new names.) In 1936, two particles were discovered that were called *mu-mesons*. They both had equal masses but opposite electrical charges. They were called mesons because their mass was 210 times that of the electron and this just happened to be about midway between the lightest and heaviest particles known at the time. (Meson means intermediate one.) The name *mu-meson* was later shortened to *muon* and was assigned the Greek letter μ, mu.

Not long after, a slightly more massive meson was discovered, the *pi-meson*. And yes, as you may suspect, this name was shortened as well to *pion*, represented by the Greek letter for pi, π. Neither muons or pions last for very long. The muon has a half-life of 2.19 microseconds. (A microsecond is a millionth of a second.) A pion was also discovered to be a messenger particle. As you will soon see, the four forces we discussed earlier all depend on what are called messenger particles that "carry" the force between interactions.

Mindwarps

Between the 1930s and late 1950s, many particles were discovered that left physicists with no idea as to their role or purpose in particle interaction. They knew that they existed for some purpose, but it was like trying to complete a jigsaw puzzle with some pieces missing. The role that many of these particles ended up having was to predict the theoretical existence of other particles that were needed to complete the picture. As more and more particles were discovered using accelerators, the particle universe began to get quite confusing. Rather than a beautiful, simple, elegant system of interactions, hundreds of particles made a rather strange picture. Could the universe really be this complicated? Fortunately things did calm down, at least a little with the theory of the existence of quarks. You'll read about those guys very soon.

From Natural to Man-Made

The cosmic ray experiments had revealed that there were many previously unsuspected particles in nature, but this natural supply of particles had its limitations. For one thing, you would have to sit around and wait until the particles you wanted happened along. All of these new particles were unstable, so if you wanted to see what these particles were like, you needed a way to produce them in large enough quantities to study. This is the reason that particle accelerators were developed. To take a normal particle and give it high energy requires that the particle be accelerated. And this is exactly what accelerators do.

There are two general classes of accelerators. In the one kind, particles are accelerated while they travel down a long straight tube. This is a linear accelerator. In the other type, the particle is made to move in a circular path by applying a magnetic field and then boosting its energy each time it comes past a given point on the circle. This sort of machine is called a cyclotron or a synchrotron, depending on how the magnetic field is applied. The first accelerators were linear and were built back in the late 1920s and '30s. They were capable of producing energies between 400,000 and 750,000 eV. Today the largest circular accelerator, which is four miles in circumference, is located at Fermilab outside of Chicago. This machine is capable of producing 1 TeV, one trillion electron volts.

The Particle Zoo

One of the most difficult aspects of particle physics is how to impose a sense of order on all of the elementary particles that have been discovered. With only a few pages left in this chapter I hope that is something I can do for you without leaving you wondering what the heck it's all about. So for sake of clarity I'm going to give you a list of some of the ways all of these pieces in the microcosmic puzzle of elementary particles can be classified. By the time we get to the end we'll see if we can come away with some simple descriptions.

➤ **Classification by interaction: leptons and hadrons**   There are three types of interactions that can affect an elementary particle: electromagnetism, the strong force, or the weak force. The electron, the muon, and the neutrino aren't part of the strong interaction at all. They are involved with the weak interaction and have been put into a small group of particles called *leptons*. All of the other particles discovered, and there are literally hundreds, are involved one way or the other with the strong interaction. This huge set of particles is made up of *hadrons*. The photon is usually put in a class by itself since it only mediates the electromagnetic interaction.

Universal Constants

A **lepton** is a particle that is involved with the weak interaction and sometimes the electromagnetic. All charged particles are affected by electromagnetism. Lepton is the Greek word for "small," and the electron is the most well known lepton.

A **hadron** is any particle that is involved with the strong interaction. It's another Greek word and it means "strong." Hundreds of hadrons have been discovered.

➤ **Classification by decay product: mesons and baryons**   All hadrons will eventually decay into a collection of stable particles—the proton, electron, photon, and neutrino. Particles in which a proton appears in the end product (the decay process often occurs as a cascade, similar to the ones created by cosmic rays entering our atmosphere) are called *baryons*. And particles whose final collection of particles is made up entirely of leptons and photons (no protons) are called *mesons*.

Universal Constants

A **baryon** (heavy one) has a proton as an end product in its decay. All baryons are hadrons and the most common ones are the proton and neutron, which make up most of the mass of ordinary atoms. For this reason everyday matter is sometimes called "baryonic matter."

A **meson** is any particle that has leptons and photons, but no protons, as the final decay particles. Remember that the muon was originally called the mu-meson, so this is a new definition of the word. They are all hadrons as well and are therefore involved in the strong interaction, too.

➤ **Classification by internal dynamics: spin**   Another classification for particles is the direction of rotation it has around its axis. It is really defined in terms of its angular momentum. Spin is defined directionally by up, down, or sideways.

➤ **Classification by electrical charge: isospin**   These particles are defined by two quantities, their spin direction and their electrical charge. The combination of these two gives a new quantity called isospin.

➤ **Classification by speed of decay: strange vs. nonstrange**   Some particles take longer to decay than others, so this allows another means to classify them. Strange particles decay in times on the order of 10^{-10} second and nonstrange particles decay in 10^{-23} seconds.

There are other methods of classification we could consider, but I think you get the idea of how it can be done. And to put it simply, we can say that all matter is made up of two kinds of particles, hadrons or leptons. But there's a little more to our puzzle that we need to complete the picture, the funny things called quarks and the particles that carry the four fundamental forces.

The Last Train to Quarkville

The periodic table was a triumph in science for organizing the elements of nature into an understandable system. And one of the powerful by-products of its development was the ability to predict elements that hadn't been discovered to fill the holes in the table. In 1961, particle physicists set up a similar table called the *eightfold way* to classify elementary particles. The eightfold way got its name because the groups of particles with associated properties naturally fit into groups of eight, called octets. And as with the periodic table, the eightfold way showed up gaps in the pattern, leading to the prediction that there must be particles as yet undiscovered.

Mindwarps

The eightfold way was deliberately chosen as a name because of the numerical association (eight) with the Buddhist "Eightfold Path or Way" to enlightenment. (Not that enlightenment was the goal of the periodic table, at least not in the strictly Buddhist sense.) Yuval Ne'eman, one of the physicists who proposed this scheme, was disappointed because he was unable to make the particles fit a pattern based on the six-pointed Star of David.

At the time that the periodic table was put together, the structure for the layout was based on associated properties, but no one really knew why the table had developed the pattern it did. It wasn't until the discovery of the atom and its basic constituents, the electron, proton, and neutron that the pattern could be explained. The same analogy can be applied to the eightfold way. Why did the particles arrange themselves in patterns of eight? The answer was that there were more fundamental particles that accounte,d for the pattern. These particles came to be called *quarks*.

And to make a long story short, it was discovered that quarks are the most basic building blocks out of which hadrons are constructed. So in a nutshell, all matter is made up of quarks or leptons. And to differentiate between their characteristics, physicists assigned rather whimsical names for them. Quarks come in six varieties of flavor: up, down, strange, charm, bottom or top, and three varieties of color: red, green, and blue, giving us a total combination of 18 quarks.

Universal Constants

Quarks are the fundamental building blocks of all hadrons, which is one of two forms that matter comes in. The word itself comes from a line in James Joyce's novel *Finnegan's Wake:* "Three quarks for Muster Mark." Also, quark is a type of German cheese. This is another typical example of physicist humor, which is probably a by-product of thinking about this strange world of particle physics.

And last but not least are the particles that "carry" the four fundamental forces. These messenger particles are called *gauge bosons*. These particles mediate the interactions of the four forces and can be broken down like this:

➤ The photon carries the electromagnetic force in particle interaction.

➤ The gluons, yes very similar to glue, carry the strong force in particle interaction.

➤ The W and Z particles carry the weak force in particle interaction.

➤ The graviton, a theoretical particle, accounts for the interaction of gravity. This particle has the least proof of existence, but it fits the picture.

And so after all of that we end up with the standard model that contains: 18 quarks, 6 leptons, and 12 gauge bosons, all of the known basic parts of the microcosm. Who's to say what else particle physics will find out about this strange universe of elementary particles. In the next chapter, we'll look at the standard model of the macrocosm, the big bang.

The Least You Need to Know

➤ The electron was the first subatomic particle discovered.

➤ Physicists measure the energy of subatomic particles in electron, or eV.

➤ Only the electron and proton can exist for indefinite amounts of time without decaying into other particles.

➤ All particles that make up matter have antimatter counterparts.

➤ The four fundamental forces of the universe are electromagnetism, gravity, the strong force, and the weak force.

➤ All matter is composed of two basic groups of particles: leptons and quarks.

Chapter 17

Scientific Origins of the Universe

In This Chapter

➤ It's either eternity or creation in time

➤ The relationship between society and cosmology

➤ Einstein's cosmological assumptions

➤ Stages of the big bang

➤ It's the best model for now

As far back as recorded history goes, there have been two sets of opposing ideas, beliefs, theories, or teachings about the origin of the universe. It has either existed eternally with no beginning or end, or it was created at some point in time and will eventually come to an end. In the first part of this book we examined the early cultural, religious, and somewhat philosophical views of how the universe began. We've also spent a little time looking at some ideas about our own beginnings from a religious and scientific point of view. In this chapter, we're going to take a brief excursion through the various theories that science has put forth to explain the origin of the universe.

By far the most popular theory in science today is the big bang theory, the idea that the universe came into existence at a certain point in time roughly 15 to 20 billion years ago. In the last 25 years this theory has moved to the forefront of cosmology. You'll meet some of the key figures whose theories have laid the foundation for the big bang. However, as you'll see as we move through this chapter, this theory is not only a product of science but also of the times in which we live. And although science would like to consider itself removed from outside influences, it can't help but be affected by the people who work in the field.

The Cosmological Pendulum

I don't think I have to reiterate for you again the two major ways in which the study of cosmology can be approached, I'm sure you remember what they are. (But just in case, you can always go back and read the opening paragraph to Chapter 1, "Once Upon a Time.") In our present day, these two methods have manifested, and in some cases crystallized into two distinct areas of science: experimentation and mathematical theory. Theorists often have nothing to do with actual experimentation and the same can be said of experimenters. And it is this distinction that has been a source of disagreement between various scientific groups who put forth one view of the origin of the universe over another. To see exactly what I'm talking about, let's trace the development of the big bang theory through its various stages. Along the way you'll get a chance to meet an opposing theory, and examine some of the reasons why the big bang was developed in the first place.

Science as a methodology likes to see itself as a revealer of the true nature of the universe, as sort of a seer that can look beneath the veil of appearance. Yet science is practiced by scientists, human beings who bring with themselves a whole set of predispositions, values and beliefs. And as in any cross section of our society, some will be seriously invested in their positions and viewpoints, taking themselves rather seriously and purporting the "correctness" of their views. Of course, there are as many who don't take this stance and seek to move beyond any personal attachment to who they are and what they've discovered.

Black Holes

It has happened more than once in the history of science that because of the fame of an individual's contribution, theory, or breakthrough in some area of science, they push their views over those of other lesser-known individuals. This happens regardless of the correctness of their explanation. Isaac Newton refused to listen to rebuttals against some of his ideas, as did Neils Bohr and even Albert Einstein. Somewhere along the way, the search for truth in science gets lost in the person identifying with the "truth" of their ideas.

Much of the history of cosmology and its theories are a reflection of these types of people and the cultures they lived in. Often the most widely accepted theory becomes exactly that, because of the forceful personality behind the ideas. And while science tries to remain free of influence from things outside of it, the scientists who practice it are still a product of the culture and the times in which they live. In other words, in relation to the theories in cosmology, whether the universe has always existed or began with a bang, can't be separated from the influence of the *zeitgeist,* or spirit of the times. While there isn't enough time to go back through history in detail and show you how the cosmological pendulum has swung from one theory to the other, I can give you a rough outline and a few examples of some time periods in which this occurred. Just remember that there are always many factors impacting how any specific paradigm develops.

➤ In ancient Greece the two basic concepts of the empirical (observation and practical application) and deductive (theoretical and mathematical) methods were intimately linked to the conflict between free citizens and the slave populace. The empirical system developed alongside the free craftsman and traders, while the deductive method, which can disregard observation and practical application, arose with the slave master's disdain for manual labor.

Universal Constants

Zeitgeist is a German word that means literally "the spirit of the times." It can also refer to a trend of thought and feeling during a period. It describes the general mood of a culture or society based on one or many influences coming from science, religion, art, politics, or even economics.

➤ The Ptolemaic system (covered in Chapter 5, "You Mean We're Not at the Center of the Universe?") was strongly influenced by the deductive method (theory and math as opposed to observation). Also at this time, we find the introduction of today's central theme in cosmology, the origin of the universe out of nothing. This ideology was developed out of the somewhat pessimistic and authoritarian worldviews of two founding Church Fathers, Tertullian and St. Augustine. The doctrine of creation *ex nihilo* served as the basis for a religious social system that saw the world as decaying from a perfect beginning to an ignominious end.

➤ During the rise of science, two central concepts of medieval cosmology were overthrown—the idea of a decaying universe, finite in space and time, and the belief that the world could be known through reason and authority. The deductive, finite Ptolemaic system was replaced with the empirical, eternal, and infinite universe that was evolving by natural processes. It was a universe knowable by observation and experiment. The triumph of science was linked to the

213

overthrow of the feudal system, out of which developed free labor and a society of merchants, craftsmen, and free peasants who questioned authoritarian power—religious, political, and economic.

➤ Today's view of cosmology is much closer to the systems of Ptolemy and Augustine than Galileo and Kepler. The big bang universe is a finite one that will eventually end in either the big chill or the big crunch, (we'll examine both of these theories in Part 5, "Supersymmetry, Superstrings, and Holograms") which like the medieval cosmos is finite in time. The universe of popular cosmology is the product of a single unique event, dissimilar from anything else that has ever occurred—just as the medieval universe was seen as a product of creation.

Universal Constants

Ex nihilo is a Latin term that translated means "out of nothing." It was an idea presented by St. Augustine that became Church doctrine later on. It is his philosophical explanation of how God created everything out of nothing, which interestingly enough can be applied to the big bang as well. Where did everything contained in the big bang come from and why did it bang in the first place?

And finally just to show you how what I outlined above can be revealed in the lives of the people living at some of those times, here are a few quotes from some famous people.

What makes God comprehensible is that he cannot be comprehended.

—Tertullian, c. 200 C.E.

If I can't laugh in heaven, I don't want to go there.

—Martin Luther, c. 1460

Religion teaches men how to go to heaven, not how the heavens go.

—Galilei Galileo, c. 1630

The most incomprehensible thing about the universe is that it is comprehensible.

—Albert Einstein, 1935

The more the universe seems comprehensible, the more it also seems pointless.

—Steven Weinberg, 1977

We may now be near the end of the search for the ultimate laws of nature.

—Stephen Hawking, 1988

Cosmonotes

Although the connections between changes in cosmology and the changes reflected in the concurrent society may be a little difficult to see, they do reflect a much deeper relationship to ideas and how people live their lives. In recent years the interaction between once separate areas of study are showing us a much bigger picture of our understanding of cosmology and other areas of human endeavor. An interdisciplinary approach has brought together philosophers, spiritual leaders, sociologists, scientists, psychologists, and cultural anthropologists to study the evolution of human ideas and to show that how we interpret and perceive the universe changes as we gain a more fundamental understanding of ourselves.

Bang That Drum

So how did we end up with the version of the big bang that we have with us today? It actually occurred in a few stages, each new version superceding the next, but still within the context of a singular event. The big bang theory is a product of the twentieth century. And in a moment we'll follow the stages that have given us the popular version. Let's compare where we were a few centuries ago to what has happened in this century, but again just generally.

The cosmology of the seventeenth and eighteenth centuries reflected a world of unlimited progress—industrial, economic, and political revolutions. It was an infinite universe open to an infinite future that reflected an advancing society. But the world of the twentieth century, where advance was halted and World Wars, along with concentration camps, atomic bombs, and political assassinations dominated the attention of society, it's not too much of surprise to see the ideas and theories of a decaying, finite cosmos rear their heads again. Now this is not to say that there haven't been some very positive things that have occurred during this time. But when the big bang theory came into being, society was going through the events just mentioned. And it has entrenched itself since then.

At the heart of the big bang is the notion that the universe is an embodiment of pre-existing mathematical laws. It doesn't begin with observation but with mathematics derived from unquestionable assumptions. And if observations conflict with theory, new concepts are introduced to perpetuate the theory. There are only a few cases of observed phenomena that support the big bang, but they're instances that have been interpreted to support the theory rather than to question it.

Mindwarps

The first idea of the big bang didn't come from science but from the literary pen of Edgar Allan Poe. Besides his creepy stories that many of us are familiar with, he was a popularizer of science and astronomy. In his essay "Eureka," he put forth the idea that God exploded the universe into creation, expanding it up to a certain point and then collapsing back on itself. He rejected the idea of an infinite universe, and reasoned instead that it was governed by gravity and would eventually contract back on itself.

Einstein's Closed Universe

For the idea of the big bang to take shape, a change had to occur in how the boundaries of the universe were defined. A few years after Einstein introduced his theory of general relativity, he put forth a cosmological view that had tremendous impact. He speculated that the universe was finite, a closed four-dimensional sphere, curved by the forces of gravity predicted in his theory. It was a static, unchanging universe governed by his elegant equations.

In 1919, the year that he had announced his views, World War I had just ended. People were recovering from the ravages of war. And progress, instead of leading to advancement, had led to death and destruction. A finite, unchanging universe was an appealing and reassuring idea from the famous man who the world liked more and more. So with that formulation the ground was laid for the development of the big bang. And contained within this view was another aspect that would have significant repercussions down through the cosmological theory of today. Einstein assumed that the universe as a whole was homogeneous, that matter is, on the largest scale, spread smoothly throughout space. You know that in general relativity, the larger the mass of an object the more it warps, or curves space. If the universe had the same density everywhere, in other words, smooth and homogeneous, all of the mass of the universe would curve space around onto itself, creating a finite sphere. But by 1919, there was sufficient evidence to support the fact that the universe was *not* homogeneous, but clumpy. This didn't bother Einstein. For philosophical and aesthetic reasons, a homogeneous universe worked better. And this precedent, to allow assumptions contrary to observation, with the idea the assumptions would eventually prove to be correct, led to that process being perpetuated down through our cosmology of today.

Black Holes

During the process of developing his cosmological theories, Einstein introduced something that he later considered "the biggest blunder of his life." In order to preserve his conception of a static universe, he added an equation called the *cosmological constant*. This mathematical term balanced the gravitational forces that were working to collapse the universe back together by introducing a repulsive force that kept it in a state of equilibrium. So the static nature would be preserved. Remember the point I made in the introduction that just because you believe something is true doesn't mean it really is? Einstein found this out firsthand.

The Primeval Atom

Without going into more detail about the development of a finite universe, let's look at the next stage in the birth of the big bang. The first version of the big bang was developed by a Belgian Catholic priest, George-Henri Lemaître (1894–1966). He had studied many of the observations made by astronomers, especially those of Edwin Hubble and Carl Wirtz. They had proposed that the universe was expanding based on the degree of *redshift* they had observed. Lemaître synthesized this information into a mathematical theory that showed two things:

➤ Einstein's idea of a static, unchanging universe couldn't be true, if it was expanding. However, he did agree with Einstein that it was finite.

➤ If the universe was finite in time that meant that it also had to be finite in space. And if the universe was expanding outward that meant that it had to have started from a point of nonexpansion, a single point in time and space called a *singularity*.

This single point that Lemaître proposed he called the "primeval atom." Remember the cascade effect of cosmic rays as they entered and collided with other particles in earth's atmosphere? This was his fireworks theory of the expansion of the universe. When the primeval atom exploded, it split up into smaller and smaller units, cosmic

217

subatomic particles that became galaxies, with those decaying into suns and solar systems. As a matter of fact, he proposed this theory for explaining the existence of cosmic rays, and cosmic rays therefore proved his theory. Of course, most scientists scoffed at his idea. It had too many flaws in its fundamental hypothesis and many of the aspects of this theory just could not be philosophically accepted by most other scientists. So this first version of the big bang died.

Universal Constants

As a galaxy or star travels farther away from us, its light shifts to the red end of the light spectrum, similar to when a train whistle's pitch drops as it passes. (That's called the Doppler effect.) When light from a distant galaxy is put through a prism, the spectrum produced shows the change in frequency of the light wave, just as when you hear the pitch drop in the Doppler effect. This change in frequency, or **redshift,** indicates that the light source is moving away from us at a high velocity.

A **singularity** is a term given to the nature of the universe before the big bang. It is a theoretical single point that has no size, and has the characteristics of being infinitely small and infinitely dense. There are no laws of physics that can explain exactly what it is.

Another Bang

With the advent of the Second World War and the development of the atomic bomb, the next version of the big bang came into existence. George Gamow, one of the Manhattan Project scientists, became the man who would push his view of the origin of the universe into the forefront of science. Upon seeing the explosion of the atomic bomb, he drew an analogy to the beginning of the universe. If the A-bomb can, in a hundred-millionth of a second, create elements still detected years later, why couldn't a huge explosion at the beginning of time produce all of the elements we have today? If the universe did come from a single point, using the equations from general relativity, Gamow theorized that the nuclear reactions created during the explosion would create all of the light elements like hydrogen and helium. And eventually as the universe continued to cool the heavier elements would be produced as well. By making some adjustments to the mathematics that explains the density of the matter in the universe, he was able to produce data that agreed pretty close to what was observed.

With theory in hand, George went on to popularize and publicize his ideas to the postwar population. Science writers and the general public quickly
embraced his theory because it was easy to understand the analogy of atomic bomb explosion. And
as the popularity of the theory increased, it became taken more as fact rather than theory. Of
course this was helped along by the publication of
Gamow's book, *One, Two, Three, Infinity,* in which
he presented the big bang as fact in the last chapter.

It wasn't long before scientists and theologians
began discussing the similarities between the big
bang and the creation of the universe in the Bible.
In 1951, Pope Pius XII made one of the first official statements of the Catholic Church regarding
the big bang theory. He stated that, "Scientists are
beginning to find the fingers of God in the creation of the universe." And if you look at the cartoon on this page you can get another idea of how
God's finger may have been used.

Mindwarps

When George Gamow's theory
was first presented, those in disagreement with it dubbed it "the
big bang." Fred Hoyle, the physicist who later proposed the
steady state theory, coined the
phrase. So the name that was
originally used to poke fun at
the idea has become the name
in the accepted cosmology of
today. Sort of ironic, isn't it?

Non Sequitur.

*(© 2001 Wiley Miller. Dist. by Universal Press Syndicate. Reprinted with
permission. All rights reserved.)*

A Big Bang Alternative

While the big bang was making its way into public consciousness, there were other
physicists that still didn't go for it. Fred Hoyle, Thomas Gold, and H. Bondi, three
other prominent scientists, put forth a theory that has come to be known as the
steady state theory. Like the big bang theory of Lemaître, it was based on philosophical premises, not scientific ones; but unlike the big bang, it proposed an eternal universe, not one that was created in time. This theory was based on the *cosmological
principle,* which contained aspects first put forth by Einstein. Basically the idea was

that the universe is homogeneous (smooth) and looks the same from any place in the universe (isotropic). If the big bang occurred, it would look different to observers at different times. The steady state theory proposed a perfect cosmological principle. In other words, the universe looked the same to all observers in all times and at all places.

Universal Constants

The **cosmological principle** is an idea in which the universe, on a very large scale, looks the same to all observers at all times and in any place. Matter is spread evenly throughout, which reflects a smooth, nonclumpy universe. It is a philosophical idea originally proposed by Albert Einstein.

Gold and Bondi suggested a unique solution for their smooth, uniformly dense universe, the spontaneous and continuous creation of matter. In each area of space, about 100 meters square, once a year a new atom comes into existence. Throughout the vast regions of space this small amount of matter would accumulate over time and maintain a constant density in an expanding universe. As old stars and galaxies die, new ones are born and formed by the constant regeneration of matter that is introduced through this spontaneous process.

This theory accounted for the creation of all of the elements in the same way that Gamow's big bang did, so it served as a good argument against the popularity of the big bang. But it never gained the scientific support that the big bang did. It was also very difficult to prove. While it went a long way in providing a sound alternative to the big bang, no one had ever observed the creation of these atoms that were supposed to be occurring spontaneously out in space. But for over fifteen years the two theories were hotly debated in scientific circles, although because of Gamow's push to popularize his, it definitely got a lot more publicity.

Microwaves to the Rescue

The most important piece of evidence in support of the big bang came in 1965 when two researchers at Bell Labs in New Jersey, Arno Penzias and Robert Wilson, verified the existence of something that was predicted by the big bang. This something is called cosmic microwave background radiation. It's sort of the diluted afterglow of the titanic explosion of the big bang. As you know, all radiation can be described by its spectrum. If you plot that spectrum on a graph it will show you how much power the radiation has at various frequencies. The big bang theory predicts that the cosmic background radiation should be in thermal equilibrium, that is, the spectrum of an object in thermal equilibrium neither absorbs nor gives up heat to its surroundings. If the source of the radiation being measured is an explosion, like the big bang, which involves the entire universe, it must be in equilibrium because there are no surroundings to get energy from or give it to.

With this confirmation of the predicted radiation, scientists became convinced that the big bang was "the answer." Papers poured in by the hundreds, all developing mathematical theories and reasons in support of its correctness. If there was any doubt, or if some aspect of the theory ran into problems, instead of questioning it, more theories were developed to plug the holes. Scientists received grants and built their reputations on the big bang theory, so no one was about to question its rightness. Besides, it is the best theory around, so why go looking for something new when what you have works pretty well. Is it cosmological laziness or just the fact that too much time, money, and energy have been put into a theory that has been presented as fact? And the assumption that Einstein introduced, that eventually a theory can be proven totally correct, is still a very strong influence, albeit somewhat unconsciously.

However, as more time has gone by, and with more experiments and stronger telescopes to verify big bang cosmology, the very thing that the experiments hoped to show, has ended up causing more problems for it. These are significant questions that the big bang has trouble answering. In the next chapter we'll take a look at exactly what the problems are and see how science thinks the universe is going to end. You'll also get a chance to learn about some new theories that may offer an alternative to the big bang.

Mindwarps

When the experiment was first set up to measure the microwave background radiation, the scientists puzzled over the excess noise that their system was picking up. There shouldn't be that much radiation. After adjusting the instruments and the antenna, they still had too much noise. After almost giving up, they found a large amount of bird droppings (who would have thought it could get that deep) inside the microwave antenna that was distorting the information coming in. Any form of excrement gives off heat as it breaks down, and that minuscule amount of heat was affecting the microwave radiation spectrum.

The Least You Need to Know

➤ Over the centuries cosmology has swung back and forth between two basic ideas: The universe has existed forever, or it began at some point in time.

➤ Science is as much a product of theory as it is of experimentation.

➤ The first theory of the big bang was put forth by the Belgian Catholic priest George Henri-Lemaître.

➤ George Gamow popularized his theory of the big bang after the Second World War and it became the foundation for what was to become the most accepted theory of today.

➤ The steady state theory was the main theory proposed to refute the big bang theory.

➤ Cosmic microwave background radiation is one of the strongest pieces of evidence in support of the big bang.

Part 5

Supersymmetry, Superstrings, and Holograms

What's string theory all about, or supersymmetry? These are some of the hottest theories in cosmology today. Physicists are seeking the elusive answer to a formula or theorem that's hopefully going to show the elegant simplicity and unity of the cosmos. And these latest theories show a lot of promise. Some physicists believe that we're very close to finding this ultimate theory. We're going to spend the next four chapters looking at what these latest hypotheses in science have to offer and to give you a clear understanding of what these theories are about.

We'll also be looking at what science thinks about how the universe will end. And besides looking at some of the most prominent theories, we'll also delve briefly into holographic cosmology, too. Not all prominent theories are necessarily the only solutions to understanding the structure of the universe. I think you'll find some of the lesser-known ones just as interesting, maybe even more so.

Chapter 18

The Accelerating Universe

In This Chapter

➤ Some problems with the big bang

➤ The essence of plasma cosmology

➤ The three fates of the universe

➤ The geometrical shapes of the cosmos

We'll begin this chapter by picking up were we left off, completing our analysis of the big bang theory. Here you'll get a chance to see the two or three significant obstacles that prevent complete acceptance of it. Be that as it may, since it is the most popular theory right now, we'll also look at the consequences of how the universe will end, if the big bang theory is the correct one. It's not a happy picture and raises a number of philosophical questions about the meaning of human existence. But again, the view of how the cosmos will eventually end is a projection of a theory, not the way it will necessarily end.

Within the context of this chapter, we'll also examine another theory that has been put forth over the last decade, the theory of plasma cosmology. It supplies answers to some of the questions that the big bang has problems dealing with. And after all of that, we'll look at the three possible fates of the universe and the geometrical shape that coincides with what cosmology knows about the structure of the cosmos at the beginning of the twenty-first century.

It's Bigger Than We Thought

There are two basic assumptions in conventional cosmology that new observations have posed problems for:

➤ The universe is, at the largest scale, smooth and homogeneous.

➤ This smooth universe is dominated by gravity alone and therefore must either contract to or expand from a single point, a singularity.

Universal Constants

In 1986, Brent Tully discovered that almost all galaxies within a distance of a billion light years of the Earth are concentrated into huge ribbons of matter that are called **superclusters.** These clusters are about a billion miles long, 300 light-years wide, and 100 million light-years thick.

However, our universe is anything but smooth, it's clumpy. But what are these clumps? They turn out to be galaxies grouped together in vast *supercluster* complexes. These are huge ribbons of matter a billion light years long. And these clumps would not warp all of space or cause it to expand or contract. Each of these superclusters would just dimple the space around it.

The idea of homogeneity has always been a problem for the big bang because for decades astronomers have known that the universe is not smooth. The usual answer to account for this clumpiness has been that even though the universe started out smooth, there were very tiny clumps in its early period. And through gravitational attraction these clumps gradually grew bigger and bigger, forming the stars, galaxies, and clusters that we have today.

The only problem with this is that the bigger the clump, the more time it takes for it to form. The age of the universe has been determined to be between 15 and 20 billion years old. Cosmologists realized that it would have taken much longer for these superclusters to form, longer than 20 billion years. It works like this. By observing the redshift (we covered that in the last chapter) of galaxies, astronomers can calculate two things: how far away the galaxies are and how fast they are moving relative to one another. And as it turns out, galaxies very seldom move faster than 1,000 kilometers per second, or one three hundredth the speed of light.

What this means is that in the 20 billion years since the big bang, a galaxy could have only moved about 65 million light years. Well, in order for these huge clusters to form, the matter contained in them would have moved at least 270 million light years, which would have taken at least 80 billion years, or four times as long as allowed by the big bang. But wait—there's more. Because the matter contained in these galaxies would first have to accelerate up to speed and the seed mass located in these regions of space would also have to form to attract matter these big distances, it would take at least 100 billion years for this all to happen. So the 20 billion years that the big bang estimates is far too short a period of time for the universe to have formed into the way it is right now.

Plasma Cosmology

We still have a lot of material to cover in this chapter, and I want to give you a brief overview of an interesting alternative to conventional cosmology. So I'm going summarize for you these two challenges to the big bang and include a third problem as well. The test of any scientific theory is based upon the relationship between its predictions and observations. Let's see how well the big bang has done:

➤ It predicts that there should be no object older than 20 billion years and larger than 150 million light years across. And as we've discussed, that's certainly not the case.

➤ It predicts that the universe, on the large scale that it exists, should be smooth and homogeneous. It's not—it's clumpy!

➤ The third problem has to do with the strongest evidence in support of it, cosmic microwave background radiation. In order for the universe to produce the galaxies we see around us, the fluctuations found in the background radiation indicates that there must be a hundred times more *dark matter* than visible matter. But there is no experimental or observable evidence that dark matter exists. It's a theory to make the big bang work. So if there is no dark matter, the theory predicts that we can't have galaxies, but we live in one—the Milky Way.

So while the big bang predicts the things in the preceding list, observations have shown them to be incorrect. However, this is the accepted theory for now, and many scientists assume that it's right. To abandon it would not be easy. Few theories in science are ever left behind when there is no alternative in sight. So what are we left with? Well, there is a new alternative on the horizon. It's called *plasma* cosmology. Here's a basic idea of what it's about.

Universal Constants

Many cosmologists think that nearly 99 percent of the universe is unobservable and made of **dark matter.** The universe we do see, the stars, galaxies, and literally everything else, only constitutes about 1 or 2 percent of the total amount of matter in the universe. The rest is some strange and unknown form of matter, particles that are necessary for the big bang theory to work. Theorists realized that there is just too little matter in the universe for the gravitational forces to have created the universe in the form that it's in today. So something has to create the needed gravity, hence the theory of dark matter. This idea was introduced about 20 years ago and has since become a fundamental part of the big bang cosmology.

Plasma is regarded as a fourth phase of matter, the other three being solid, liquid, and gas. It is a hot state of matter in which electrons have been stripped from atoms to leave positively charged ions, which mingle freely with the electrons. The Northern lights are a naturally occurring form of plasma, as is St. Elmo's fire. You've probably seen the "Eye of the Storm" or similar plasma balls in stores. They're those really cool objects that have the tiny electrical storms inside the glass spheres. When you bring your hand in contact with the glass surface, the bolts of plasma electricity inside react to ions surrounding your hand.

The advocates of plasma cosmology believe that the evolution of the universe in the past must be explained in terms of the processes occurring in the universe today. In other words, events that occur in the depths of space can be explained in terms of phenomena studied in the laboratories on earth. This approach rules out the concepts of a universe that began out of nothing, somewhere in time, like the big bang. We can't recreate the initial conditions of the big bang in laboratories. The closest we can get is in the particles created in accelerators. Plasma cosmology supports the idea that because we see an evolving universe that is constantly changing, this universe has always existed and has always evolved, and will continue to exist and evolve for eternity.

Another aspect of this new theory is that, while the big bang sees the universe in terms of gravity alone, the plasma universe is formed and controlled by electricity and magnetism, not just gravitation. With the introduction of electromagnetism the "clumpiness" of the universe and the fluctuations in microwave background radiation can be easily accounted for. Even the expansion of the universe can be explained by the electromagnetic interaction of matter and antimatter.

Cosmonotes

Since all that is being provided for you is a simple summary and basic explanation of plasma cosmology I would recommend that you check out the list of recommended reading in this area in the Appendix B, "Suggested Reading List." There is a lot more to this theory than I can elaborate on in the space of a few pages, so if you're interested in finding out more about these new ideas, I suggest you look into some of the books I've recommended. There is still very little support for this theory because the big bang is the one that many believe is the correct interpretation of the origin of the universe, and to question the validity of this theory is not on the minds of many of today's cosmologists.

And while electromagnetism forms the basis for plasma cosmology, it is also the basis for our technological society that surrounds us today. Plasma technology has stimulated research for better computer screens, how radio and radar transmission can be increased, and may be the answer to developing the long-sought-after genie in the bottle: fusion energy. So in the long run it holds the possibility of not only providing a better description of the origin and structure of the universe, but it can also lead to a whole new area of advanced technology. When we get to Part 6, "Old Endings and New Beginnings," I'll be discussing some material on the Eastern traditions and their approach to cosmology. There are some interesting correlations between their understanding of the universe and the ideas behind plasma cosmology.

The Standard Model

Given the state of cosmology today it's probably a good idea to spend the rest of this chapter providing you with an understanding of where the conventional theories are taking us. The standard model in cosmology is the big bang. And while plasma cosmology is a potential alternative, it'll still be a while before it finds more adherents. So let's continue with some aspects of the big bang that we haven't covered yet. The focus of this discussion revolves around the three possible scenarios of how the universe will end. Two of these can be summarized as follows:

➤ The big crunch is the mirror reverse of the big bang. This would happen if the amount of gravitational force in the universe were strong enough to bring the currently expanding universe to an end. The universe would then stop expanding and eventually collapse back in on itself.

➤ The big chill is the opposite of the big crunch. In this scenario the universe would continue to expand forever. There is not enough gravitational force to stop the expansion, and eventually all of the stars and galaxies would exhaust their supply of energy and the universe would cool down and there would be no heat left. Everything would be cold, cold, cold. Brrhhhh!

I think you can see from the preceding summary that the most important factor in determining the fate of the universe is gravity. And to give you a little better idea of how these two fates are possible, let me give you a couple of examples to illustrate how they would work. If you toss a ball in the air, it will reach some maximum height where it stops momentarily and then returns to your hands. In this case the force of gravity is strong enough to brake and decelerate the moving ball to the point where it reverses its velocity. This is what the big crunch is like.

If, on the other hand, we launch a rocket into the far reaches of outer space, never to return, we have an example of the big chill. In this case, the rocket's energy of motion exceeds the gravitational energy, and the rocket continues on forever because the Earth's gravity was incapable of stopping the motion. So in both examples it's the amount of gravity that affects the outcome. But there are some other factors we need to look at besides gravity. One is the *second law of thermodynamics* and the other is the shape of the universe.

This second law is also linked to another physical process: entropy. Entropy is the measure of the amount of disorder in the universe. It reflects the concept that once something wears down it won't build itself back up without outside help. A broken egg will never reassemble itself. In a closed system, entropy never decreases, only increases or remains constant. So if the universe is a closed system, it will slowly run down and die. If on the other hand it is an open system, order can increase and entropy decrease, because there is input coming from somewhere else that will help bring this about.

Universal Constants

The **second law of thermodynamics** simply states that things wear out. But to be a little more scientific, it is one of three fundamental laws in the study of heat. It expresses the physical law that heat cannot flow from a cold object to a hotter object of its own volition. For example, an ice cube in a cup of warm water melts as the heat flows into it from the water, ending up with a cup of water slightly cooler than the one we started with. But you never see ice cubes spontaneously forming in cups of water; they can only be formed by pumping the heat out of the water, which is what happens in your freezer. The process can only go in one direction. It's not reversible.

Cosmonotes

The study of what goes on inside systems is called general systems theory. It's a fascinating area of study that examines the rules governing closed and open systems, which are the two types of possible systems. These systems are found all around us in nature, society, the Earth, the solar system, the universe, and even ourselves. We've learned a lot about the interdependence of systems and how the imbalance in one can affect the balance of others. A good example of this is the impact we have had on our environment. What do you think is out of balance in us, which has in turn put the environment out of balance?

The Alpha and the Omega

Now that we know that gravity, entropy, and the second law of thermodynamics all play a role in defining the fate of the universe in big bang theory, let's put it all together and look at the three possible fates of the universe. I've already mentioned two—the big chill and the big crunch—but there is a third alternative as well. Let's get a little more into the role of the scientist and view this whole process and all three possible scenarios in ideas and concepts that cosmologists use.

231

When discussing the gravitational energy of the universe as a whole, there is a direct correlation to the mass density of the universe: the higher the mass density of the universe or an object, the greater the gravitational energy. For example, a *neutron star* is a very compact star that is created when the core of a very massive star collapses. A neutron star can have a mass similar to that of the sun, but a radius that is seventy thousand times smaller. Consequently, gravity near the surface of the neutron star is about five billion times stronger than near the surface of the Sun.

The gravitational energy of the universe is precisely equal to its kinetic energy (the energy an object possesses by virtue of its motion) for a particular value of the mass density in the universe. This value, which separates eternal expansion from eventual contraction, is called the *critical density*. If the density in the universe is higher than the critical density, gravity will prevail; the expansion will stop and contraction will occur. If the density is lower than the critical density, the universe will continue to expand forever. The third scenario is a borderline state between the big crunch and the big chill. In this case, when the kinetic energy exactly equals the gravitational energy (in other words, when the mass density is exactly equal to the critical density) the expansion still proceeds forever, but the speed at which the universe is expanding approaches zero as time progresses.

In physics, as you've seen before, a Greek letter is used to denote this ratio of the actual density to the critical density. The letter used is omega, Ω. Omega is the twenty-fourth and final letter of the Greek alphabet, an appropriate symbol in this case. So using this letter as cosmologists do when they talk about how the universe will end, eventual contraction and the big crunch correspond to a value of omega larger than one. Or in other words, the ratio of the mass density to the critical density is greater than one. If the universe expands forever for eternity, the value of omega will be less than one. This means that the ratio of the mass density to the critical density will be less than one. And in a universe that expands forever, but has a speed that slowly approaches zero, omega will be exactly equal to one. In this case, the ratio of the mass density to the critical density is equal.

Universal Constants

A **neutron star** is made almost entirely of neutrons, contains roughly as much mass as our sun, but is packed into a sphere only about 10 km across, with the density of an atomic nucleus. If a star has more than three times the mass of our sun, at the end of its life, it will collapse still further from a neutron star into a black hole, as gravity overwhelms all quantum effects and matter is crushed out of existence.

Mindwarps

The methods used to find the value of omega are very similar to those used by the great Carthaginian general, Hannibal. Before his battle with the Romans in Cannae, in 216 B.C.E., he wanted to accurately determine the size of the Roman army. He did this by gathering exact information on the amount of food supplies that were furnished to the Roman army. By calculating how much food each soldier consumed, he was able to pretty accurately determine the size of the army. Similarly, in calculating the amount of cosmic mass density, scientists rely on observations that this density produces, the gravitational attraction.

The Need for Dark Matter

Therefore, in order to answer the question about the ultimate fate of the universe, we need to determine the present value of omega, or in other words, determine whether the density of the mass in our universe is higher than, lower than, or equal to the critical value. It sounds simple, doesn't it? Well this is where support for the concept of dark matter comes in. It seems that without this hypothetical mass there just isn't enough matter in the universe to account for the gravitational forces that are holding the galaxies and superclusters together, so the universe would go on expanding forever. Without the presence of dark matter, the overall density of the universe (mass density) is 100 times smaller than the critical density, which would give us a value of omega less than 0.01. That figure is too small and doesn't agree with other values for omega that have been estimated.

There are several methods that have been employed by astronomers to determine omega. I'll give you an idea of how one of them works. From the speeds of gaseous clouds around the center of individual galaxies and the speeds of galaxies in clusters and superclusters, astronomers have established that the dark matter overweighs the luminous matter by a factor of 10 or more. So, the

Universal Constants

The **critical density** is a mathematical value assigned to the relationship between the gravitational energy of the universe and the mass density of the universe. It is used to determine the possible fate of the universe as well as the shape of the universe.

value of omega inferred from the gravitational dynamics in clusters and superclusters is about 0.2 to 0.3. This is not the official estimate however. As it turns out, omega has been inferred to be equal to one, and the lower values, it is thought, only represent an inability in the methods used to uncover all the dark matter that exists. The theory that was developed in support of an omega value of one is rather complicated and beyond the introductory scope of this book, but it's important to note that even before the theory was developed, physicists expressed a strong prejudice favoring omega equals one, simply on the basis of *aesthetics*.

Universal Constants

In philosophy, **aesthetics** is the study of beauty. In physics it is the deep belief shared by many physicists that the theories of the universe must be beautiful. Of course, there's nothing in the laws of physics that requires this, it is a human bias that has been a "guiding light" in discovering the underlying structure of cosmology. As humans we have a tendency to imbue the cosmos with a sense of elegance and beauty, mathematically, musically, and structurally. Almost every theory we've discussed in this book has that as a predominant feature. But what if it isn't? Could we be happy living in an "ugly" universe?

Living in the Flatlands

Gravity, the force that is the key ingredient in big bang cosmology is, as you've learned, also used to define the value of omega. Gravity is also employed to define the shape of the universe. As you'll see in a moment, there is a direct correlation between the shape of the universe and the value assigned to omega. In a universe that is closed, the value of omega is larger than one. In this model, the mass density is sufficiently high that gravity would stop the expansion and the universe will collapse back into itself. Geometrically, this corresponds to space-time with a spherical shape, in other words a big round ball. The mass density causes the space to curve back on itself. In such a universe, if you travel along a straight line, (which would really be a great circle) you would eventually return to the point from which you started. There are other strange features of this shape for a universe: Parallel lines eventually cross each other, the shortest distance between two points is not a straight line but a curve, and the sum of the three angles in a triangle is always more than 180 degrees. (There are always only 180 degrees in a triangle on a flat surface. And when airplanes fly great distances on the Earth, they never fly in a straight line; it's shorter to follow the upward or downward curve of the Earth.)

The second possibility is a universe that's open. This corresponds to a value of omega that is smaller than one. In this case the gravitational field is too weak to stop the expansion and the universe will expand forever. The geometrical shape of this type of universe is the opposite of the previous model. Instead of space-time curving back on itself and creating a finite volume, space curves away from itself in an open universe, which produces an infinite space. The shape can best be described as looking like a saddle. And of course, it would also have the opposite features of the sphere. Parallel lines would eventually diverge, the sum of the angles of a triangle would always be less than 180 degrees, and the shortest distance between two points is a *hyperbola*.

Our third possibility is that of a universe in which omega is precisely equal to 1. This, as you know, puts the universe on the borderline between eternal expansion and eventual collapse. It expands forever, but the speed at which it expands becomes closer and closer to 0. The geometric shape of this universe is flat. Yes, flat, just like a tabletop or wall. And in this case we have our familiar features: parallel lines stay parallel, there are always 180 degrees in the sum of the angles in a triangle, and the shortest distance between two points is a straight line. Does the fact that our everyday world reflects the geometry of flat space have anything to do with it being flat? Not really, it just makes it easier to understand.

It's Out of Control

This then is the state of big bang cosmology today. The general consensus is that omega is equal to one and that the shape of the universe is flat. So what does this mean? What have astronomers and cosmologists discovered about what's happening in the cosmos to reflect this? Well one of the most significant things that appears to be happening right now is that the speed at which the universe is expanding is accelerating. The universe is flying apart at ever increasing speeds. Right now, the 100 billion galaxies that we can see with our telescopes will recede out of range one by one. (Of course, we're talking about hundreds of millions of years before this happens.) Tens of billions of years from now the Milky Way will be the only galaxy we'll be able to see. But by then our sun will have shrunk to a white dwarf (a very compact star close to the end of its life and which emits white light) that will be providing very little light and heat to who or whatever is left on earth. But in cosmological time, billions of years is very small. The universe will continue to exist; each of the stars and suns in all of the galaxies will slowly burn out. By

Universal Constants

A **hyperbola** is a curve formed by cutting a cone with a plane that is more steeply inclined to the base than it is to the side of the cone. So what does that mean in English? It's just a special type of curve that is essentially U-shaped.

235

the time the universe is one trillion trillion trillion trillion trillion years old, all the dead matter will collapse into black holes, and these will eventually disintegrate into stray particles occasionally running into each other. So what was once a universe lit by countless stars and galaxies will end up being a vast, empty, dark and unimaginably lonely place. And with that cheerful thought we'll bring this chapter to a close. The kinds of questions that this scientific outlook raises in religious and philosophical domains will be addressed in Part 6.

The Least You Need to Know

➤ The big bang theory is the standard model of the universe.

➤ Plasma cosmology is a possible alternative to the big bang theory because it can account for problems that can't be explained by the big bang.

➤ The universe can end in the big crunch, the big chill, or expand forever, but slowing as it does so.

➤ The geometric shape of the universe is closed, open, or flat.

➤ Most scientists who adhere to the standard model currently believe that the universe is flat.

Chapter 19

Supersymmetry, Superpartners, and Superman

In This Chapter

➤ The search for unity

➤ The problem with unification

➤ Quantum units of measurement

➤ Matching particles and extra dimensions

➤ The super in supersymmetry

A couple of centuries from now, historians of science are going to look at the twentieth century as the point in time when science originated the central idea that the universe is in a constant state of change and is always evolving. Although this may not sound like any big deal, it does express a new way that science expresses its cosmology. Of course whether it remains this way, only time and new theories will tell. But I think it's safe to say that there won't be a return to the idea that the universe is static and unchanging. However, how and what the universe is evolving toward is another question and raises an almost unlimited number of questions. We'll come back to those questions in a few chapters. I just wanted to whet your appetite.

But to get back to the original statement, that science sees the universe as changing and evolving reflects a level of confidence in the scientific paradigm that is due to the knowledge that has been brought together from the two areas of cosmology that we've been examining in the last 10 chapters: the microcosm and the macrocosm. Particle physics (microcosm) and astrophysics (macrocosm) have been able to develop a better understanding of the universe because of what each has learned from the other. It's one of the reasons why I devoted a couple of chapters to explaining the world of the atom. And it's the unification of these two areas of cosmology that is the Holy Grail that science seeks. For when the microcosm is united with the macro-cosm, the most complete picture of the structure of the universe will have been at-tained, or so it is thought. The discussion in this chapter and the next will also be within the context of the standard model, since all the knowledge gained works from that theory.

The First Three Minutes

To understand the search for the ultimate theory (or as you already know, the TOE), we need to know what already has been accomplished. This won't take very long, just the first three minutes of the creation of the universe. It is during this period of time that the microcosmic forces that shape our universe today had their origin. Given the point in time where we are now, physicists can mathematically calculate the expan-sion of the universe in reverse. It's how they've been able to estimate how old the uni-verse is. It also provides a way to restructure the universe to the point of the big bang. From that beginning, it is then possible to work forward again and see how the uni-verse unfolded.

Black Holes

While it's easy to visualize the explosion of the big bang as analogous to an atom bomb, it was really much different than that. A common misconception is to think that it occurred at a specific point in space, but there was no space for it to explode into. It's much more like blowing up a balloon. The space inside continues to expand the more air you blow into it. And that's what the expansion of the universe was like. It wasn't expanding into space that was already there. It was creating space and time as it expanded. There was and is no physical object that exists outside of the space/time created by the big bang.

By the time the universe was just three minutes old, the four fundamental forces were already in existence and the basic building blocks for helium and hydrogen, the two elements that make up 99 percent of all visible matter, were present. The key feature of the explosion of the big bang was heat. For it was only after the universe began to cool that the forces and particles began to differentiate. And we're talking 10^{32} de-grees. Temperatures so hot that nothing that exists today even comes close. It's one of the difficulties in trying to recreate the conditions under which the four forces separated and particles came into existence. One of the uses of particle accelerators is to try to get as close as possible to the energy in these early moments as possible. But accelerators would have to be built that are trillions of times more powerful than anything we currently have to match these conditions. However, they have provided some of the initial conditions from which further knowledge of the early universe can be extrapolated.

Attempts at Unification

At one time in the history of the universe, the four fundamental forces, along with the particles that carry these forces, and all the myriad elementary particles we covered in Chapter 16, "Forces, Particles, and Some Cosmological Glue," came from a singularity. The big bang is the ultimate unifier, the one thing from which everything else has come. So the search for a theory that unifies everything back to this one point might reveal the reason why it happened in the first place. This is one possible reason why the search is on. Or is it simply reductionism, the method that science is known for? It could also be that sense of a search for beauty and elegance that we discussed in Chapter 18, "The Accelerating Universe," that drives physicists and cosmologists to find the theory that will unify everything. Whatever the reason, it lays at the core of present-day cosmology.

Einstein devoted a large part of the last thirty years of his life in an attempt to unify electromagnetism and gravity. He wanted to show that these two forces were different manifestations of the same force. His valiant attempts proved fruitless because he hadn't known about the other two forces, the weak force and the strong force. So without the knowledge of these other forces, any attempt at unification would of course fail. But even in today's search for unification, trying to combine gravity with the other three forces is what still eludes solution. As we'll see shortly, gravity is the main force that poses the most problems for unification.

Cosmonotes

One of the best descriptions of the early universe is Steven Weinberg's book *The First Three Minutes*. Although written more than 25 years ago, it's still one of the best introductions to the whole field of particle physics as well as a good in-depth look at the earliest moments after the big bang.

There have been some remarkable achievements toward unification of the forces. In the late 1960s, physicists Steven Weinberg, Abdus Salam, and Sheldon Glascow showed that the nuclear weak force and the electromagnetic force were different aspects of one force, now called the electroweak force. Since then, particle accelerators have been able to recreate the energies needed to combine these two forces. When the universe was about 10^{-11} seconds old, particles had kinetic energies of about 100 times the *rest-mass energy* of the proton.

Since such energies can be created in particle accelerators, the theory has been confirmed by experimentation. The carriers or messenger particles are those that carry the four forces (you learned about those in Chapter 16) and the theory linked the carrier of electromagnetism, the photon, with three previously undiscovered particles, the carriers of the weak force. The existence of these predicted particles, the W^+, the W^-, and the Z^0, were finally discovered in 1983.

Universal Constants

You remember Einstein's most famous equation, right? It is this equation that allows physicists to convert the mass of a particle into an equivalent amount of energy. This also enables them to know how much energy is needed to recreate those particles. So the **rest-mass energy** of a particle is simply the energy obtained from a particle when its mass is converted to energy.

The general idea behind unification is very simple, regardless of the difficulty encountered in trying to make it happen. All the basic interactions were unified in earlier stages of the universe when it was, as you know, much, much hotter. The universe was highly *symmetrical,* in the sense that interchanging any of the forces or particles among themselves would have resulted in no change. This understanding of symmetry has led to trying to unite the four forces under the theories of supersymmetry, a theory that we'll also be covering shortly. But as the universe expanded and cooled down, at certain critical temperatures, symmetry breaks occurred, and eventually the four interactions gained their distinct identities. These certain critical temperatures are called *phase transition temperatures,* and knowing exactly what the degree of heat is at these temperatures is another key ingredient in knowing when the four forces differentiated. The fact that we observe four distinct forces now simply reveals the fact that the universe is already much colder than it was.

The unification of the strong nuclear force with the electroweak force occurs, according to the GUTs, or grand unified theories, only at energies of about 10^{15} times the rest-mass of a proton. That's an incredibly higher amount of energy than was required to show the connection between the electromagnetic and weak force. That was only 100 times the rest-mass of a proton, or 10^2.

Particles in the universe had such energies as theorized by the GUT's when the universe was less than 10^{-35} seconds old and the temperature was above 10^{28} degrees. The high energies that are required to unite the strong force and the electroweak force can't be achieved by any accelerator. Current accelerators can peak out at 1TeV, or one trillion electron volts. These energies have been created at the four-mile circular accelerator at Fermilab, but even that level of energy is only 10^{12}. Another three powers of 10 need to be produced, and unfortunately, each power of 10 becomes incredibly harder to attain.

Universal Constants

Symmetry is a property of a system that does not change when the system is transformed in some manner. For example, a sphere is rotationally symmetrical because it doesn't change when it is rotated on its axis. There are many different kinds of symmetry. Mirror symmetry, for example, is exactly what is says. An object has an identical symmetrical equivalent—it's just the mirror image of it. And something that is asymmetrical has no uniformity, regardless of how it is transformed or moved.

Phase transition temperature is the point where matter changes from one state into another. Solid, liquid, and gas are the most common, while plasma is a rare fourth state of matter. One of the more familiar phase transitions is water into ice and vice versa. The phase transition temperature in this case is 32^0F or 0^0C.

Quantum Mechanics vs. General Relativity

Before we go on, I think it's a good idea to clearly present the problem of unification. What is the main obstacle in the way of uniting the four forces and all of the elementary particles? Well, remember that what physicists have been trying to accomplish is the uniting of the microcosm and the macrocosm. These two areas of cosmology are represented respectively by quantum mechanics and general relativity. In the case of quantum mechanics, we have a world that operates on uncertainty, probability, and complimentarity. If we could look through a microscope at this tiny universe, we would see random quantum undulations resembling something looking like a storm on the North Atlantic Ocean. And if we took this into outer space, into the smooth gravitational field of a planet, you would no longer have the smooth warp of space-time described by the spatial geometry of general relativity. At this microscopic level,

Universal Constants

Quantum foam is the term used by physicists to describe the violent activity of the quantum world. When you combine all of the characteristics of quantum mechanics—such as wave functions, probability, and uncertainty—you get a pretty active interaction among the forces and particles. You can also think of it as a quantum soup, happily bubbling away.

the gravitational field would be warped by the frenetic energy of the *quantum foam*. So this fundamental incompatibility of quantum mechanics and general relativity occurs not on the level of everyday life, or even in the vastness of the universe, but at the most fundamental level where the building blocks of matter have their existence.

The Gravity of the Situation

The inability to reconcile general relativity with quantum mechanics didn't just occur to physicists. It was actually after many other successful theories had already been developed that gravity was recognized as the elusive force. The first attempt at unifying relativity and quantum mechanics took place when special relativity was merged with electromagnetism. This created the theory of *quantum electrodynamics,* or *QED.* It is an example of what has come to be known as *relativistic quantum field theory,* or just *quantum field theory.* QED is considered by most physicists to be the most precise theory of natural phenomena ever developed.

Universal Constants

Quantum electrodynamics, or **QED,** is the theory that describes the way electrically charged particles interact with one another and with magnetic fields through the exchange of photons. Also known as **relativistic quantum field theory,** it's quantum because it includes all of the quantum characteristics like probability and uncertainty; it's a field theory because it includes Maxwell's electromagnetic field equations; and it's relativistic because it incorporates the concepts of space and time from the special theory of relativity.

In the 1960s and '70s, the success of QED prompted other physicists to try an analogous approach to unifying the weak, the strong, and the gravitational forces. Out of these discoveries came another set of theories that merged the strong and weak forces called *quantum chromodynamics,* or *QCD,* and quantum electroweak theory, or simply the electroweak theory, which you've already been introduced to.

If you examine the forces and particles that have been combined in the theories we just covered, you'll notice that the obvious force missing is that of gravity. But hope is around the corner! The search for the *primary theory* is still underway. And in order to understand the theories that hold the highest possibility of resolving the dilemma, we need to examine and define the small little corner of the universe where this unification could take place. Although we've spent a lot of time discussing forces and particles, theories and solutions, the one thing we haven't discussed very much are the units of measurement that will allow physicists and you to understand and communicate their discoveries. In other words, we need a clear picture of the scale at which unification takes place.

The Planck Scale

In other chapters we've talked about how important measurement is to comprehending the "relative" nature of the universe. We use common terms like inches, feet, meters, pounds, and kilograms to describe physical characteristics such as weight, length, or distance of objects. What kind of units of measurement would we use to describe our essential unifying theory? Well, the best place to start is with units that are common to all aspects of nature. Universal constants interestingly enough fit the bill. Let's see what they are.

➤ Planck's constant, h: This constant defines the ratio between a particle's energy and its frequency. If used in conjunction with the mass and the charge of an electron, the size of all atoms, of all kinds, anywhere in the universe can be determined.

➤ The speed of light, c: This speed is constant under all conditions, and is one that you're already familiar with.

➤ Newton's constant, G: This constant measures the strength of the gravitational force. Einstein proved that energy and mass are convertible into one another and gravitation is a force proportional to the amount of energy (mass) a system has, so everything in the universe feels the gravitational force.

Universal Constants

Quantum chromodynamics, or **QCD,** is the theory that describes the way quarks interact with one another by the exchange of gluons (remember gluons are the messenger particles that carry the strong force). Quarks come in three different colors: red, blue, and green. The "chromo" part of the name comes from the way the "color" of the quarks changes when it interacts with gluons.

Universal Constants

Some physicists don't like the phrase, "the theory of everything," to describe the ultimate theory of the natural world. It's somewhat misleading, because it's not really the theory about everything in nature. It doesn't include the weather, baseballs, psychology, or people. They feel that the idea of a "final theory" or **primary theory** sounds more appropriate.

With these three constants, we should be able to combine them into units of measurement that will reflect the scale at which the unification takes place. Each of these units is named after Max Planck and can be described as follows:

➤ **Planck length** $(hG \div c^3)^{1/2}$ This is the quantum of length, the smallest measurement of length that has meaning. It's equal to 10^{-35} meter and is about 10^{-20} times the size of a proton.

➤ **Planck time** $(hG \div c^5)^{1/2}$ This is the quantum of time, the smallest measurement of time that any meaning. Within the framework of the laws of physics as understood today, we can say that the universe came into existence when it had an age of 10^{-43} seconds.

➤ **Planck mass** $(hc \div G)^{1/2}$ This mass is equivalent to 10^{-5} grams. This is small by everyday standards, but 10^{19} times the mass of a proton, and would be contained in a volume roughly 10^{-60} times that of a proton. This represents an enormous density that has not occurred naturally since the big bang.

Black Holes

Just when science thought it was safe to assume that universal constants, like the speed of light, were absolute, a discovery comes along to throw a wrench in the works. Recently a team of international researchers found something that could make the basic laws of nature questionable. While examining the light coming from a quasar, which is an extremely bright object that produces 10 trillion times the energy per second as our sun, scientists found discrepancies in the spectrum patterns they were comparing. The differences suggested that something, possibly the speed of light, had changed by the time it reached the earth, trillions of miles from the quasar. To quote one of the scientists, "We don't know what has changed, we don't know whether it's the speed of light, or the electron charge or Planck's constant," and he added, "There are theoretical reasons for preferring the speed of light." In two or three years they'll know whether they're right or wrong. If they're right, that will have a profound impact on physics.

These three units, along with others such as Planck density and Planck temperature, define the Planck scale. And all of these units express the smallest possible measurement that can be made in trying to understand what happened in the infinitesimal moments after the big bang. When we discuss superstring theory in Chapter 20, "It's

All Held Together with Strings," the units we just discussed will be used to describe the level at which unification is possible.

Symmetry Breaking

Concepts of symmetry are very important in physics. Space-time symmetries are all around us in the everyday world: the right and left sides of an animal body, the circular disk of the Sun and Moon, a wallpaper pattern, or even a repeated musical rhythm. And symmetry has become a fundamental way of expressing the laws of physics. If an object has symmetry, it has fewer features to describe and contains less information. We need to describe the theme and the number of ways in which it is repeated, like one house in a row of identical houses. Scientific laws can be put in these terms: Something is the same at all times and all places. Now on the other side of the coin, if a system undergoes a phase transition to a more differentiated state, it has more features to describe and also contains more information. For example, water looks the same in all directions, but when it changes and becomes a snowflake, it becomes more complex. A snowflake crystal looks the same in only six directions. This is an example of *symmetry breaking*.

Universal Constants

Symmetry breaking is an important process in both biological evolution and the evolution of the universe as it moves in time through different eras. Wherever structure becomes more complex, symmetry, or at least the original symmetry, is lost. It is this breakdown of symmetry that theorists are trying to understand in reverse, because at the beginning of the universe, it was perfectly symmetrical; and as it cooled, symmetry breaking took place, creating the more complex and varied world of particle physics.

Symmetry breaking is equally important in the world of particle physics. It is really the cornerstone concept out of which the theory of supersymmetry was born. Physicists believe that just after the big bang, all of the forces of nature were identical and all elementary particles were the same. But within an instant, symmetry was broken. First the color force between quarks broke away from the electroweak force, and hadrons developed very different masses from leptons. Next, the electroweak force fragmented into two parts—electromagnetism and the weak force ... and on and on. I think you get the idea. I don't want to confuse you with further broken symmetries.

The Spin Doctors

An important feature of the quantum world and an important bridge between the microcosm and the macrocosm is the connection between the *spin* of elementary particles and their statistics. It appears as though nature divides the quantum world into two classes of particles. The fermions, matter particles that include all leptons and quarks, are associated with fractional spins, such as $1/2$, and bosons, the particles that carry the four forces and have a whole number spin, such as 1 or 2. Each of these two divisions has an accompanying statistical system that helps to define their energy states, which is important to know so that physicists can relate the spin of the particle to its location and how it's interacting with other particles.

Universal Constants

All of the elementary particles have a property called **spin.** It's a hard concept to define because it can only be described within the framework of quantum mechanics. However, to suit our purposes you can think of it as rotation around an axis, like a spinning ball, or the rotation of the Earth.

One of the strangest features of quantum spin is shown by the behavior of fermions also know as "spin $1/2$ particles." If an object like the earth turns in space through 360 degrees, it returns to where it started. But if a spin $1/2$ particle rotates through 360 degrees, it arrives at a quantum state that is measurably different from its starting state. To get back to where it started, it has to rotate through another 360 degrees, making 720 degrees, a double rotation. One way of picturing this is that the quantum particle "sees" the universe differently from how we see it. What we see if we turn through 360 degrees twice are two identical copies of the universe, but the quantum particle is able to discern a difference between the two copies of the universe. This weird notion of a particle having to rotate through another 360 degrees plays an important role in the theory of supersymmetry, as you'll soon see.

Introducing SUSY

I grew up reading stories about Superman and Batman during the Golden Age of comics when these superheroes fought villains and beings not only from outer space and here on earth, but from other dimensions as well. This early introduction to visualizing more than just the four dimensions of space-time left me with a desire to find out if these other dimensions were really real or just science fiction or fantasy. In the last 25 years the concepts of other dimensions have become a mainstay in movies and television that deal with these same categories. Words such as subspace, warpdrive, hyperdrive, hyperspace, warpspace, multiverse, multidimensional, and superspace are familiar terms to followers of science fiction. But the concept of other dimensions, like many aspects that were once only part of science fiction, has become part of mainstream science. In both superstring and supersymmetry, additional dimensions play a key role in the theory.

In the 1970s, the success of *gauge symmetry* in understanding the interaction between forces and particles encouraged theorists to attempt to find a geometrical description of everything in terms of one great symmetry, or supersymmetry. This supersymmetry, or SUSY for short, utilizes two main concepts to unify the four forces and all of the elementary particles:

➤ The addition of four other dimensions to the already existing four dimensions of space-time to give us a total of eight dimensions.

➤ Each real world particle has a corresponding superpartner that interacts with it and which mediates the movement from four dimensions to eight dimensions and vice versa.

Universal Constants

Gauge symmetry is a concept used in field theory (like in an electromagnetic field) to describe a field for which the equations describing the field do not change when some operation is applied to all particles everywhere in space. Remember that in science, symmetry is a property that means it's the same at all times and all places. The term "gauge" simply means to measure. The point is that fields with gauge symmetry can be remeasured from different places without affecting their properties.

Interdimensional Travel

Let's look at these two features in a little more detail. As mentioned earlier, nature has seemingly divided herself into two main classifications of particles, matter particles or fermions and force particles or bosons. In geometric terms, the key difference between these two kinds of particles is the spin rotation. In the previous discussion on spin, we saw that fermions need 720 degrees or two rotations in order to get back to where it started. But bosons have to rotate only once or just 360 degrees to get back to where it started. SUSY is a kind of symmetry that unites these two different patterns of behavior in one geometric framework.

Supersymmetry works by attaching another four dimensions to the four dimensions of ordinary space-time. The resulting eight-dimensional geometry, which is known as superspace, provides the needed room for the extra rotation that a fermion needs to get back to its original starting configuration. However, these extra dimensions are not space or time dimensions and are different than the dimensions I'll be talking about in the next chapter on string theory. The extra dimensions are mathematical/ geometrical constructs that utilize the mathematics of SUSY to rotate particles in and out of the extra four dimensions. Let me explain.

Cosmonotes

The assigned values of $1/2$ spin or 1 spin reflect the relationship that particles have to 360-degree rotation. In other words, a value of 1 means that it takes 1 complete rotation of the particle to complete 360 degrees. And just in case you didn't know, there are 360 degrees in a circle. The fermions, which only have a spin of $1/2$, means that even though they rotate 360 degrees, which is one complete rotation in the normal sense, they are only "halfway" to the point where they first started from. They have to complete another whole rotation, or 720 degrees altogether to get back to the point where they started.

In the mathematics of SUSY, there is an operation that is equivalent to rotation in the everyday world. But instead of rotating an object in four-dimensional space-time, this operation rotates an object from the usual four-dimensional space-time into the eight-dimensional geometry inhabited by fermions. And of course, there is an equivalent operation that rotates an object out of eight-dimensional geometry inhabited by fermions back into the everyday geometry four-dimensional space-time. What this means is that it is possible to transform bosons into fermions and fermions into bosons, and what we see as two different kinds of particles is an illusion created by geometry.

No, I'm Sparticles!

If this is a little confusing, I hope to make it less so by the time we're through. In our discussion above, rotating fermions in this way would not produce any of the known bosons. Within the structure of this theory, none of the known bosons correspond to the rotated versions of known fermions, and none of the known fermions correspond to rotated versions of known bosons. So if supersymmetry is to apply to the real world, there must be a supersymmetric particle, also called a superpartner, or a sparticle, for every known type of boson and every known type of fermion. This of course doubles the number of varieties of particles in the world. The following is a table of some of the known particles and their corresponding sparticles.

Mindwarps

Besides the predictions in supersymmetry for all of the superpartners, there is a prediction for another particle related to the "mechanism" that is responsible for the symmetry breaking. The currently favored mechanism is known as the *Higgs field*, after Peter Higgs of the University of Edinburgh. The Higgs field acts as a sort of "party pooper" in that when it achieves its lowest energy state (which all systems like to achieve), it breaks the symmetry. The prediction is that the Higgs field is carried by a massive particle, known as the Higgs Boson, sometimes affectionately called the "God particle." The discovery of this particle is currently one of the goals of the huge accelerators. If found, this particle would open the doors to revealing how symmetry was first broken shortly after the big bang and lead to a firm foundation upon which can be built the "primary" theory.

Known Particles That Transmit Forces and Their Possible Superpartners

Name	Spin	Superpartner	Spin
Graviton	2	Gravitino	$\frac{3}{2}$
Photon	1	Photino	$\frac{1}{2}$
Gluon	1	Gluino	$\frac{1}{2}$
$W^{+,-}$	1	$Wino^{+,-}$	$\frac{1}{2}$
Z^0	1	Zino	$\frac{1}{2}$
Higgs	0	Higgsino	$\frac{1}{2}$

Known Particles That Make Up Matter and Their Possible Superpartners

Name	Spin	Superpartner	Spin
Electron	$\frac{1}{2}$	Selectron	0
Muon	$\frac{1}{2}$	Smuon	0
Tau	$\frac{1}{2}$	Stau	0
Neutrino	$\frac{1}{2}$	Sneutrino	0
Quark	$\frac{1}{2}$	Squark	0

And as with other broken symmetries in particle physics, the implication is that at higher energies, or in other words, at distances much closer to the point of expansion, there was a complete symmetry. In this case, the complete symmetry would show that each type of boson was accompanied by a fermion superpartner and vice versa. The reason why the symmetry is broken is assumed to be that the supersymmetric partners are much more massive than the counterparts we know today, and could be manufactured only at very high energies. So the search is on to find these massive particles in the most powerful particle accelerators we have.

To put this theory in a nutshell, if that's really possible, it could go something like this. Supersymmetry is the idea, or hypothesis, that the equations of the TOE will remain unchanged even if fermions are replaced by bosons, and vice versa. The replacement occurs in the equations, not for fermions or bosons in the real world. But by showing how these particles interact, even just mathematically, gravity, along with the other three forces, because of the particles that carry them, becomes unified with the matter particles. The broken symmetry, which reflects the structure of the everyday world around us, is again unified under the geometrical/mathematical–particle/sparticle model of supersymmetry.

The Least You Need to Know

➤ Heat is the key ingredient in the earliest moments after the big bang—making symmetry possible.

➤ The electroweak force unified the electromagnetic and the weak force—the first accomplishment in the uniting of two of the four fundamental forces.

➤ The main obstacle to developing a theory of everything is the incompatibility of quantum mechanics and general relativity.

➤ The Planck scale is the closest physicists can come in measuring the qualities of time, length and mass that have a meaning in relation to the big bang.

➤ Supersymmetry requires two concepts to work: an additional four geometric dimensions and corresponding superpartners to all known particles.

➤ Supersymmetry is a purely mathematical theory; none of the predicted particles have been found, but the search is on.

Chapter 20

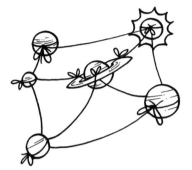

It's All Held Together with Strings

In This Chapter

➤ The evolution of string theory

➤ Superstrings and violin strings

➤ The key features of string theory

➤ More extra dimensions

➤ M-theory

Congratulations! You've made it through quantum mechanics, relativity, and supersymmetry. We finally come to the most promising of all the theories of everything, superstring theory. And while there are other theories to explain the structure of the universe, such as plasma cosmology and the yet to be discussed holographic universe, as far as pure science is concerned, superstrings offer the best solution.

In the last chapter, "Supersymmetry, Superpartners, and Superman," the two fundamental aspects of supersymmetry—extra dimensions and superpartners—laid the foundation for discussing superstring theory. For as you'll see, the latest theory of everything relies on extra dimensions as well as the union of the two pillars of physics, quantum mechanics and general relativity. And although these extra dimensions are needed, the real key to the whole theory is the strings themselves. What are they? What are they made of? What exactly do they do? These questions will be answered in due course. So let's unravel what superstring theory is all about.

A Brief History of String Theory

String theory began with the observation that elementary particle resonances (the different energies at which new elementary particles are produced in the colliding beams from particle accelerators) form regular patterns, not unlike the overtones from a plucked string. This led the Italian physicist Gabriele Veneziano to propose in 1968 that the hadrons, the strongly interacting elementary particles, are in fact energy vibrations of incredibly small strings. In geometry, the most elementary unit is a point, like the period at the end of this sentence. But Veneziano thought that the most elementary unit in geometry was not a point in space, but a tiny extended string.

His theory was pretty ingenious, but soon ran into difficulties. It was discovered that the only way for the mathematics of the theory to satisfy both quantum mechanics and general relativity was if the strings existed in a space of 26 dimensions for bosons or 10 dimensions for fermions. At the time, physicists persisted in trying to connect string theory to the theory of quarks, suggesting that quarks are actually the ends of strings. The reason that isolated quarks have never been seen became immediately obvious—break a string in half in the hope of capturing a free end and all you end up doing is creating a new end.

Cosmonotes

If you remember in the last chapter on supersymmetry, the goal was to unite the particles that carry the four forces—the bosons, with the particles of matter—the fermions. This is the route being taken in physics to unite all of the particles under one theory. The common concept to both supersymmetry and superstring theory is the need to have more dimensions in order for these particles to unify.

Mindwarps

As we've seen in earlier chapters, music has been used as a metaphor to describe the structure of the cosmos more than once. We had Pythagoras's "music of the spheres," Schrödinger's and de Broglie's vibrating string and musical atom, and now the musical analogy is again being used to describe the fundamental quality of string theory. Who's to say if in reality the universe isn't one big cosmic symphony!

Let me explain this in another way, just in case the last paragraph left you feeling a little unclear. The quarks inside hadrons are held together by the exchange of gluons, and the effect is as if two quarks are joined by a piece of elastic. The force between quarks (the color force, which also indirectly gives rise to the strong interaction) is so strong that the energy in the "elastic" is comparable to the mass energy in the quarks themselves. Under these conditions, a pair of quarks is joined by the color force and behaves in many ways like a stretched piece of string. A good image for this would be the chain shot used in sea battles in the days of sail. A pair of cannonballs joined by a chain would whirl around one another as they stuck the rigging of a man-of-war ship, and would end up doing far more damage than two single balls passing through the sails.

Cosmonotes

If the theory of superstrings, like supersymmetry, has so many dimensions, why do we sense only three of space and one of time? Theory has it that all the dimensions were created at the instant of the big bang, when the size of the entire cosmos was far smaller than that of an elementary particle. In the period of rapid expansion that followed, four of these dimensions expanded and unrolled, while the remaining dimensions remained tightly curled up. Today the four dimensions define the universe we live in, while the other dimensions are effectively invisible, yet their effects are felt throughout the forces of nature.

But to get back to our story, this first version of string theory was soon superceded by the development of supersymmetry. The basic concepts developed by string theory about the unification of fermions and bosons through extra dimensions replaced the notion of strings, and supersymmetry replaced the theory that had given it birth. Once the idea of supersymmetry had been placed in the minds of physicists, it was easy to incorporate it into the then standard model of the particle world, as we discussed in the last chapter. As a matter of fact, that is the way generations of students after 1976 were introduced to supersymmetry, without any mention of strings at all. In the early 1980s, the English physicist Michael Green and the American John Schwarz married the ideas of string theory to those of supersymmetry to create, yes you guessed it, superstrings.

The Cosmic Onion

Between 1984 and 1986, thousands of research papers were written by physicists from around the world. This three-year period is regarded as the first superstring revolution. What was this first theory about? Why was it replaced with the second revolution in superstring theory? And what are these strings made of? To answer these questions let's look at some of the basic concepts that define this first revolution.

To clarify the most essential point of string theory, we need to change how we view the whole set of elementary particles. In SST, which is the abbreviation I'll use when discussing superstring theory, the idea of a particle as just a point in space, which is the most general concept associated with a particle, is replaced with the idea of a tiny vibrating string, a string that's connected together to form a loop. That's it! You now have the secret of the theory of everything. But as with all theories, there is more to the whole picture.

What are these strings made of? In truth, no one knows, and I'll tell you why in a moment. I can tell you that they are the size of the Planck length. Remember that unit of measurement from the last chapter? That was 10^{-35} meter, or about 100 billion billion (10^{20}) times smaller than a proton. Now, regarding their composition, well that's the trick. If they are truly the most fundamental units out of which everything else is made, then to say that they are composed of something would mean that there was something even smaller than these strings, and then, of course, they wouldn't be what they are. We would have the dilemma of a cosmic onion, in which each layer that's peeled away just reveals another layer.

Mindwarps

The analogy of a cosmic onion to a continually unfolding universe is an interesting image. While many physicists think that SST will turn out to be the theory of everything, some also feel that strings may not be the ultimate building blocks. There may even be units that make up the string loops. If this is the case, who's to say that the process may not keep on going. If the universe is infinitely large as well as infinitely small, as we peel away layer by layer of its structure, we may find the cosmic onion to have no end to its layers.

Musical Strings

A deeper look at the analogy of a vibrating musical string can take us to a better understanding of how the first SST revolution saw string theory as a possible answer to unification. If we use a violin as our musical instrument to explain how strings work, I think you'll get a clear picture of how physicists understand SST. The four strings on a violin can each vibrate at almost an infinite number of vibrational patterns called *resonances*. These are the wave patterns that fit between the two fixed ends of the violin string. When we hear these different vibrational resonances, we experience them as the different musical notes. The way the strings vibrate in string theory is very similar. Each closed loop can support almost an infinite number of resonant vibrational patterns within its structure. Instead of having a string fixed on both ends, the closed loop provides the same kind of structure so that it can vibrate in the same way as the violin string.

Universal Constants

Resonance is usually understood as the oscillation of a system, like a guitar, piano, etc., at its natural frequency of vibration, triggered by an outside stimulus with an appropriate frequency. For example, if you play a note on a piano with the exact frequency of one of the open strings of a guitar, the string on the guitar will resonate, even though it hasn't been plucked. You may have also experienced this while singing in the shower. Certain notes will resonate better than others giving you a deeper- or mellower-sounding voice. In the way the word is used in the context of string theory, it simply refers to the number of waves or the frequency at which a string can vibrate.

So how does this analogy help explain the central concept of SST? What this means is that just as the different vibrational patterns of a violin string create the different sounds that we hear, so, too, the different vibrational patterns of the strings in string theory create the different masses and force charges.

The Important Part

In this section of the chapter we'll make the connection between vibrating patterns and the theory of everything. So this section is the heart of the chapter. If you understand this section, you'll have a good foundation on which to explore SST in more detail. It's worthwhile to repeat the last sentence of the last paragraph, because herein lies the key. The characteristics and properties of an elementary particle, its mass, and the forces that it can carry, are determined by the precise resonant pattern of vibration that its internal string "plays" or performs. For example, let's take a particle's mass. The energy of any vibrating string is defined by its wavelength and its amplitude. The shorter the wavelength and the greater the amplitude, the more energy it has. If we compare this to a violin again, you can see that if you pluck one of the strings vigorously (with more energy), the more intense the vibration. And if you pluck it more softly (with less energy), it vibrates less vigorously.

Mindwarps

Do you know of anywhere else in nature that uses strings as a fundamental structure? I think a moment's thought will reveal that nature has reserved the string for a special role, as a basic building block for other forms. The essential feature of life on earth is the string-like DNA molecule, which contains the complex information and coding of life itself. When building the stuff of life as well as subatomic matter, strings seem to work incredibly well. The distinguishing feature of a string is that it is one of the most compact ways of storing vast amounts of data in a way in which information can be replicated. For living things, nature uses the double strands of the DNA molecule, which unwind and form duplicate copies of each other. Also, our bodies contain billions upon billions of protein strings, formed of amino acid building blocks. Our bodies then, in a sense, can be viewed as a vast collection of strings—protein molecules draped around our bones.

As you know from special relativity, mass and energy are interchangeable. And the greater the mass, the more energy there is to convert, and the same is true in reverse. So if we calculate the energy of the vibrational pattern of the internal string of an elementary particle, we should be able to convert that information into finding its mass. Lighter particles have internal strings that vibrate less energetically (like the softer plucked string), so of course, the heavier particles have internal strings that vibrate more energetically (like the harder plucked string). As you can see, this method can be used to determine a basic characteristic, the mass. But what about finding out about the force it carries?

If you know the mass of a particle, you can also determine its gravitational properties, because as you know, there is a direct relationship between the mass of an object and how this mass reacts to gravity. This means that there is a direct association between the pattern of string vibration and the particle's response to the gravitational force. Using this reasoning, physicists have shown that the detailed aspects of a string's pattern of vibration can also be related to the three other forces. In other words, the vibrational pattern of a particular string will determine which force (weak, strong, or electromagnetic) is being carried by that string. This vibrational pattern is then also used to determine which of the particles are associated with which force. Of particular importance was the discovery that among all the different vibrational patterns, one matched perfectly with the properties of the *graviton*, the particle thought to carry the force of gravity. Here then is also the unification of gravity with the other three forces.

Universal Constants

The **graviton** is the theoretical particle assumed to carry the gravitational force. Since the other three of the four fundamental forces is carried by a particle, it is a natural conclusion that gravity should be carried by a messenger particle as well. It plays a part in quantum gravity analogous to the role of the photon in electromagnetic interactions described by QED that we covered in the last chapter.

Our key concept then can be defined in the following way—each elementary particle is composed of a single string, or better yet, each particle is a single string. And all strings are totally identical. What makes particles different is the fact that each of their respective strings has a unique vibrational pattern, their own vibratory fingerprint you could say. And because every physical event or dynamic process in the universe consists of the most basic units that interact with each other, either producing matter or applying a force, SST provides a system of unification that makes it a good candidate for a TOE.

The Second Requirement, Extra Dimensions

Okay, you now have one of the most important parts of string theory under your belt. But you're not off the hook yet. There is another important point we need to discuss and this point is equally significant. (I know, you thought I said the last section was the most important. It is, but so is this section.) We'll start by stating exactly what this other requirement is and then work backward from there. As in SUSY, extra dimensions are required for SST.

You already know that the vibrational pattern created by the strings defines the type of particle it is. But what physicists found was that these patterns needed more than our three dimensions of space to express all of their possible patterns. In a sidebar, I recently explained that after the moment of the big bang, four dimensions uncurled and many other dimensions didn't. They remained curled up. It is these uncurled dimensions that are necessary for the patterns of the strings to vibrate in. Let me reword this a different way for clarity.

All of the possible vibrational patterns that strings can assume are constrained and strongly influenced by the shape of the extra dimensions that are twisted up and curled back on each other. This means that the extra-dimensional geometry determines the fundamental physical attributes, like particle masses and charges, which we observe in the usual three large space dimensions of the everyday world. In other words, the vibrational patterns that define each type of particle are based on the shape of the extra dimensions that the string vibrates in. Got it?

Cosmonotes

The difficulty in describing the extra dimensions that are needed in SST lies in the ability to express their geometrical form. These extra dimensions are not just crumpled up any old way. In 1984, four physicists showed that the extra six dimensions needed can be described using what's called a Calabi-Yau shape. The name is derived from the two mathematicians, Eugenio Calabi and Shing-Tung Yau that developed the model. The closest I can come in telling you what it looks like is that it resembles the famous "Gordian Knot" that Alexander the Great unraveled with one swipe of his sword. Just imagine an incredibly complex ball of twine and you'll get a good visual image, though still not an accurate representation.

What do these curled up dimensions look like? That's harder to describe than anything else we've discussed so far. There are very exotic names given to the unusual forms that these curled up dimensions can take. And they're too lengthy and mathematical to explain here. But I think I can give you a good idea of how just one extra dimension is possible to imagine, so let's see what that looks like.

The Kaluza-Klein Dimension

The idea that our universe may have more dimensions than meets the eye originated in 1919 with a German mathematician by the name of Theodor Kaluza. Oscar Klein, a Swedish physicist, later refined it when he applied it to quantum theory. They introduced the idea of a five-dimensional model to unify gravity and electromagnetism. In the same way that gravity ripples space in four dimensions, the Kaluza-Klein equations showed that electromagnetism ripples in the fifth dimension. But where is this fifth dimension? The standard explanation is that the fifth dimension is hidden from us by *compactification*. Over the next couple of paragraphs you'll get a good idea of what this term means.

Cosmonotes

The term "Kalazu-Klein model" has come to be applied to any version of grand unification that operates in many dimensions and requires compactification. Superstring theory requires a total of ten dimensions, six more than the three of space and one of time. The extra dimensions are accounted for by compactification in just the same way as the one extra dimension is in the original Kaluza-Klein theory.

The most common analogy used to describe this extra dimension is that of a garden hose. If you stretch a long garden hose across a gorge, from a distance it looks one-dimensional. The actual thickness of the hose would not be evident from a distance. And if there were a bug crawling on it, and someone wanted to know where it was, you would only have to tell him its location based on one dimension, left/right. But now, if we view the hose with a powerful pair of binoculars, we see that the hose actually has thickness to it. The bug on the hose has in reality two choices facing him when he moves. He can move left/right or clockwise/counterclockwise in a circular motion around the hose. So from a distance what appeared to be one-dimensional, upon much closer examination is revealed to be two-dimensional. And this is the central point of the extra dimensions, and part of the Kaluza-Klein theory. One of the dimensions is long, extended, and easily visible, just like the four dimensions we live in all of the time. But the other dimension on the hose is short, "curled"

Mindwarps

Kaluza first proposed his theory to Einstein in 1919, who upon reading it was initially impressed, but upon careful reconsideration felt that the notion of an extra dimension was too unprovable. However, two years later, after he digested the theory more fully, he gave it his full support. And even though it's called the Kaluza-Klein theory, the two of them never worked together on it.

up and can only be discovered with much greater precision. This is the dimension that is hidden from our normal experience and is compacted into tiny, curled-up shapes.

Unification in Higher Dimensions

In the last chapter, I introduced you to the idea of "quantum foam." When we looked at the quantum universe, magnified almost to the Planck scale, we saw that the uniform smoothness of space warped by gravity was a boiling, frothing area of space far different than that created by gravity alone. So how does SST get rid of that quantum foam? If you can imagine a birthday cake that just had the icing put on it, it's rather lumpy and chaotic. But with the use of a large butter knife, you can "smear" the roughness out into a smooth surface. This is essentially what the tiny, curled-up extra dimensions do. Because of their geometric shape, they can "smear" the quantum foam, more or less, into the smoother surface reflected in the forces of gravity.

If we magnify an area of space almost to the Planck scale, we see that it is made up of tiny, curled-up shapes, the extra dimension of the Kaluza-Klein theory. Of course in their theory, there is only one extra dimension, but I think you have a good idea of how this one dimension is described. So to reiterate their theory one more time: The spatial fabric of the universe can possibly have two types of dimensionality, the large, extended, easily visible dimensions that we operate in, three of space and one of time, and the hidden, curled-up, circular dimensions. This is a new dimension, but one that exists curled up at every point in space in the familiar larger dimensions. In the same way that you can move, up/down, left/right, forward/backward in three dimensions, if you were small enough you could also move in the other dimension as well.

Mindwarps

While most of the extra dimensions in SUSY and SST are spatial dimensions, some theorists think that to arrive at a full understanding of a primary theory that will completely describe and unify the microcosm with the macrocosm, some extra time dimensions will be required as well. For example, a time dimension applied to the spin of a particle could allow it to rotate through its full revolution so that it would return to the same moment in time it was at before it began its rotation. You could think of it as a particle starring in its own version of the movie *Groundhog Day*.

How Many Dimensions Are There?

In the last chapter, we saw how the extra dimensions in supersymmetry allowed for the extra rotation that fermions required to get back to their original starting point. In superstring theory the extra dimensions play a different role. But are there eight dimensions as in SUSY, or five, as described by the Kaluza-Klein theory? Maybe there's more? Well as it turns out there are a different amount of dimensions for each of the two basic groups of particles. Do you remember what those are? The fermions make up matter, and the bosons carry the force charges.

If we begin with just one extra dimension, as in the Kaluza-Klein theory, that one dimension is a simple curled-up circle, very similar to our second dimension that we found upon close inspection of our garden hose. Each dimension that is added changes the geometric shape of the curled-up dimension. But in principle, whether it's one or ten extra dimensions, the concept of them being curled up and hidden, as opposed to being large and extended is still the same.

Universal Constants

Modular functions were first discovered by the Indian mathematical genius Srinivasa Ramanujan. He has been compared to a bursting supernova, illuminating the darkest, most profound corners of mathematics, before being tragically struck down by tuberculosis at the age of 33. Scattered throughout his notebooks are sets of very obscure equations called modular functions. These are some of the strangest equations ever found and appear only in distant, unrelated branches of mathematics. One of these modular functions, called the Ramanujan function in his honor, appears in string theory as the underlying mathematical theory that shows how strings vibrate in 10 and 26 dimensions.

So to answer our question "How many dimensions are there?" we could say, "As many as it takes to make the theory work." There could be an infinite number of dimensions. But as it turns out, at least for SST, 10 dimensions work for fermions and 26 dimensions work for bosons. Remember that a particle is defined by the particular vibrational pattern is has and that pattern is defined by the shape of the space in which it vibrates. It was found that the vibrating waves of fermions travel clockwise using 10 dimensions and the vibrating waves of bosons travel counterclockwise in 26 dimensions. This is one of the most unusual aspects of the theory and also one of the least understood. The mathematical theory of *modular functions* is where this is all explained. But no one has the slightest understanding why 10 and 26 dimensions are singled out as the dimensions in which strings vibrate.

The Short Version of the Second Superstring Revolution

Most of this chapter has been devoted to explaining the fundamental concepts found in the first superstring revolution. In this last section of the chapter, I would like to summarize for you the main points of the second revolution in string theory. The essential concepts that you learned about string theory in the previous pages still lie at the core of this second revolution, but there have been some subtle additions that are too technical to go into. So for the sake of your brain and mine, I'll just define for you what has happened.

By the late 1980s and very early '90s, string theory was definitely the theory that everyone felt would eventually lead to the TOE. There were just two basic problems with it.

➤ Physicists had developed five different versions of string theory. Without spending the next five chapters telling you all about them, I'll just give you their names. We have Type I, Type II, Type IIB, *Heterotic* O, and Heterotic E. They are all fundamentally the same, but the way that they incorporate supersymmetry and explain the vibrational patterns vary from theory to theory.

➤ These five versions each ended up having many possible solutions to the equations that defined them. There were different possible ways to curl up the extra dimensions with each solution corresponding to a universe with different properties. What lies at the heart of this problem is that the equations in string theory are so complicated that no one knows their exact form.

Universal Constants

Heterotic comes from the same Greek root as in "heterosexual," which implies the combination of at least two different things. In string theory it refers to the version that explains the two different sets of dimensions that are applied to the fermions and bosons, 10 dimensions for fermions and 26 dimensions for bosons.

However, in 1995, which is considered to be the date for the start of the second superstring revolution, some great new developments were introduced to resolve the two problems mentioned above. As it turns out, the equations that define each of the five versions are very intimately related. So rather than having five distinct versions, physicists are convinced that there is one theory that brings all five together under one umbrella. Picture a five-pointed star, like a starfish, with one of the five versions at each tip and you have the new model for SST called M-Theory. And although there is still a lot of work to be done, we can summarize two of its chief features.

➤ M-theory operates in eleven dimensions instead of ten. The extra dimension provides the space in which all five versions can be synthesized into one theory.

➤ Besides the vibrating strings that define string theory, it's been discovered that there are other vibrating objects: two-dimensional membranes, undulating three-dimensional blobs called "three-branes," and a whole bunch of other things as well.

Just one final thought before we close this chapter. Although SST gives us a compelling formulation of the theory of the universe, the fundamental problem is that an experimental test of the theory seems beyond our present day technology. In fact, the theory predicts that the unification of all forces occurs at the Planck energy, or 10^{19} billion electron volts, which is about 1 quadrillion times larger than energies currently available in our accelerators. This can be a little disappointing, because it means that experimental verification, the engine that drives progress in physics, is no longer possible with our current generation of machines or with any generation in the conceivable future. But be that as it may, the true nature of M-theory remains mysterious, which is one of the suggested meanings for the M. Acquiring a full understanding of M-theory could well be the Holy Grail of the twenty-first-century physics.

The Least You Need to Know

➤ Elementary particles are not considered particles in string theory, but are tiny vibrating loops called strings.

➤ The resonating vibratory pattern is different for each string and defines the "fingerprint" that identifies what particle it is.

➤ The resonating pattern is in turn defined by the number of dimensions and the shape of the dimensions as well.

➤ The Kaluza–Klein theory was the first one to put forth the concept of an extra dimension.

➤ M-theory is the most recent development in SST and combines the five versions of string theory into one comprehensive theory.

➤ There is no verifiable experimental evidence that the higher dimensions in strings or string theory exist.

The Universe As a Hologram

Have you ever seen a hologram? They are those eerie, ghostly, two-dimensional images that appear to be three-dimensional. The first time I saw one was in *Star Wars,* when Princess Leia's holographic image was projected by R2D2 to Luke Skywalker. They've become quite popular over the last twenty years or so. Most charge cards have a holographic image imprinted on them. And you can often go to an art show where holograms are the main works being presented.

So what do holograms have to do with theories of the universe? Well, as you'll be seeing shortly, some cosmologists think that the universe is a giant hologram. And while in this chapter we'll be looking at holographic theory from a scientific perspective, in other chapters coming up we'll also see how it relates to human consciousness. But this will bring to a close the purely scientific part of this book. In the last four chapters we'll be looking at theories of the universe from philosophic, psychological, and symbolic points of view. And contained within that discussion, we'll pursue the role of human consciousness and how important that is for any theory of the universe.

What's a Hologram?

The discussion that follows in this chapter is going to be based on the work of just a few people. Holographic cosmology, like plasma cosmology, sits in a small corner of the universe with the other theories that aren't part of mainstream physics or science. Nevertheless, it deals with fundamental anomalies that most scientists in mainstream science have chosen to ignore or have dismissed as unimportant. Most of the insightful work done in relating cosmology to holograms has been done by David Bohm, a University of London physicist, a protégé of Einstein and a widely respected quantum physicist. The other scientist that we'll look at in the next chapter is Karl Pribram, a neurophysiologist at Stanford University and author of many books on the physiology of the brain. The ideas and theories developed by these two individuals form the essential ingredients of holographic theory and how it relates to the universe and our mind.

But before we delve into those ideas, it would be good to provide you with a little background on exactly what a hologram is. In 1947, Dennis Gabor formulated the mathematical theories out of which would come the development of the hologram. At the time, he was working on improving the electron microscope. The mathematical equations that he was using were based on a type of calculus invented by an eighteenth-century Frenchman by the name of Jean Fourier. The equations he had developed were called *Fourier transforms*.

Universal Constants

A **Fourier transform** is a mathematical way of converting or transforming any simple or complex pattern into a language of simple waves. To get an idea of how this works, let's take the example of a television set and the camera in the studio. The television camera in the studio takes an image and converts it into electromagnetic frequencies. Those frequencies are then broadcast or sent via cable to your television set at home. There your TV converts those frequencies back into the images you see on your screen. And the Fourier transforms do the same thing. His equations convert images into waveforms and back again into patterns or images.

By experimenting with different kinds of film, Gabor used these equations to help him convert a picture of an object into a blurry set of interference patterns (remember those from Young's double slit experiment?) on film, and then converted those interference patterns back into an image of the original object. Of course, today the mathematics are taken out of the process, because now the whole thing is done with lasers. But his original insight into how images and waves can be converted into each other and vice versa made the development of holograms possible.

Laser beams are used to create holograms because the light emitted from a laser is a very pure form of coherent light. It's especially good at creating interference patterns. Without the invention of the laser, the holographic images we see around us today would not have been possible. To create a hologram, the single beam of light that's emitted is split into two separate beams. The first beam is bounced off the object to be photographed, while the second beam is allowed to collide with the reflected light of the first. The interference pattern that's created by the intersection of these two beams is then recorded on a piece of film. When you look at the image on the film, it doesn't look at all like the object that was photographed. It simply looks like a group of concentric rings, similar to the surface of a pond after you throw a rock in. However, when you shine another laser beam or even just a strong source of light through the film, a three-dimensional image of the object appears.

Cosmonotes

The word laser or L.A.S.E.R. is really an acronym for light amplification by stimulated emission of radiation. Inside a laser is some form of crystal, gas or other suitable substance, in which atoms, when stimulated by focused light waves, amplify and concentrate these waves and then emit them in a narrow, very intense beam of one color or frequency.

When a holographic image is projected, you can actually walk around it and view it from different angles. It looks almost solid, yet if you put your hand in it, there is nothing there. But one of the most unusual aspects of holograms is what lies at the core of the theories we're going to cover. If you take a piece of holographic film and cut it in half, each half will contain the entire image of the object photographed. And if you continue to cut the pieces in half again and again, the entire image of the object can still be seen on each piece of film.

Action at a Distance

This astounding bit of information about holographic imagery is what led both Bohm, and as we'll see in the next chapter, Pribram to develop their theories. To see how each of them developed their ideas, we need to go back and look at one or

Black Holes

The phenomenon of seeing objects for each piece of film that is cut only occurs on holographic film whose images are invisible to the naked eye, not on images that don't require special illumination. So don't cut anything in half, like your credit card, to see if this is true!

two of the weird aspects of quantum behavior. Remember that the double slit experiment is considered to be the central mystery of quantum mechanics. When the particle that is sent through the slit is observed, it behaves as a particle, but when it is not observed, it behaves as a wave. That led Neils Bohr to develop his famous Copenhagen interpretation in which the electron didn't come into existence until it was observed. Of course, Schrödinger's famous cat was the rebuttal to that argument, but that idea of observation altering the outlook of an experiment still lies as a core principle in quantum mechanics and is also the basis for Heisenberg's uncertainty principle.

Mindwarps

In the 1930s a Russian scientist Nikolai Berstein discovered that our physical movements might be encoded in our brains in a language of Fourier waveforms. He painted white dots on dancers and then filmed them walking, jumping, dancing, and other movements against a black background. When he converted their movements into a language of waveforms by using the same equations (Fourier transforms) that form the basis for holography, he found that the waveforms contained hidden patterns that allowed him to predict his subject's next movement to within a fraction of an inch.

Einstein and Bohr got into some very famous debates about these weird aspects of quantum mechanics and one of the most famous arguments put forth by Einstein is known as the EPR paradox (named after the three men who developed it—Einstein, Podolsky, and Rosen). Their argument went something like this. Because measurements disturb a quantum system, the idea was to perform measurements on two separate systems to see if they were able to have identical internal observed characteristics. In other words, two similar quantum systems are allowed to interact until their internal states become correlated. Now put the two systems on opposite sides of the room or miles apart with observers at each box. If the boxes are both opened at precisely the same time and the internal states are measured and still identical, then somehow the systems were able to communicate to each other their information instantaneously so they could know which internal states to agree upon. This, according to Einstein and the boys, would be impossible because instantaneous transfer of information is not possible since nothing can travel faster than the speed of light. So how could observation bring these characteristics into being? They thought they had him. But Bohr responded by saying that instead of believing that some faster than light mechanism was at work, he simply said that if subatomic particles don't exist until they're observed, then they could no longer be thought of as independent things. They were part of an indivisible system and could not be thought of as separate entities.

Black Holes

In the last 20 years and even more recently in the last 5 years, there have been a growing number of scientists, mostly physicists and biologists, who are questioning some of the more unusual aspects of quantum theory that have been ignored or swept under the carpet by the traditional scientific establishment. While most of science, especially physics, is content to stay within the framework of quantum mechanics developed over the past 50 years, others feel that to ignore these puzzling aspects is to miss a significant part of quantum theory that will ultimately lead to greater understanding. Some even think that the TOE will continue to elude physicists until these fundamental anomalies are addressed. And as you'll see in this chapter, the new theories that deal with these unusual aspects offer some remarkable insights into cosmology and the role we play as an essential part of the cosmos.

A Remarkable Event

In 1952, David Bohm suggested a way to test the EPR paradox, (although only as a thought experiment), and in 1964, John Bell showed how this thought experiment might be carried out in reality. Finally in 1982, the experiment was successfully carried out in Paris by Alain Aspect and his colleagues. The experiment is a little complicated to get into, but the outcome proved to be a momentous point in the history of science. Of course you didn't hear about this in newspapers or on the evening news and, unless you read science journals, you may have missed it altogether. However, the implications of this experiment have the potential to shake the very foundation of science itself and may be one of the reasons why it has been somewhat ignored.

Aspect and his team found that under certain circumstances, subatomic particles such as electrons and photons are able to instantaneously communicate with each other regardless of the distance separating them, whether it's 20 feet or 20 billion miles. Somehow each particle always seems to know what the other is doing. This, of course, violates the ultimate speed of the universe, the speed of light. It also shows that the universe has an interconnectedness that can't be explained within the context of traditional physics. The important feature of these experiments is that they have directly detected *nonlocality*. Based on this finding, there is no need to invoke the collapse of the wave function or any other interpretation of quantum mechanics or indeed to accept quantum theory at all. What has been revealed as a fundamental truth about the universe is that there are correlations that take place instantaneously, regardless of the separation of the objects involved.

Universal Constants

Nonlocality is a term used to describe the way in which the behavior of a quantum entity, like an electron, is affected by not only what is going on at one point (the "locality" of the entity), but also by events that are going on at other places (other localities), which can in principle be on the other side of the universe. These nonlocal influences occur instantaneously as some form of communication which Einstein called a "spooky action at a distance, not just faster than the speed of light but infinitely fast."

Nonlocality

Why is nonlocality such a big deal? You would think that a discovery of this sort would attract a lot of attention. Well, that's just it. The ramifications of the results found in experiments dealing with nonlocality would require restructuring our understanding of reality from the point of view of quantum mechanics. Let me explain why. Classical physics and much of quantum theory states that physical reality is local—a point in space cannot influence another point beyond a relatively short distance. Besides the Aspect experiments, in 1997 other experiments were conducted in which light particles (photons) that originated under certain conditions were sent in opposite directions to detectors located about seven miles apart. As in the Aspect experiment, the results indicated that the photons interacted or communicated with one another instantaneously. And even though seven miles is not a large distance to us, in relation to the quantum universe, it would be the same as halfway across the universe.

The most fundamental implication of nonlocality is that since every particle in the universe has been "entangled" with other particles, like the particles in the experiments, physical reality on the most basic level is an undivided wholeness. It also demonstrates that physical processes are vastly interdependent and interactive, an organic whole, in some ways very similar to Plato's ancient cosmology (we covered that back in Chapter 2, "The Music of the Spheres"). There are also implications for us in human terms, for there is no longer the need for accepting as fact the stark division between our minds and the natural world that has preoccupied much of Western thought since the seventeenth century and the development of Cartesian duality. As you'll see in some of the material in this chapter and in ones yet to come, human consciousness can be viewed as emergent from and seamlessly connected with the entire cosmos.

Mindwarps

The implications of nonlocality extend into the realm of biology as well as physics. Recent studies on the evolution of the human brain indicate that the logical foundations of mathematics and ordinary language are much more similar than previously imagined. This understanding leads to knowledge that reveals a much deeper and more intimate connection between our minds and the natural world. In the upcoming chapters you'll get a chance to see just how deep this connection may possibly be.

Quantum Potential

We'll come back to nonlocality in a little while. You'll see that it plays an important part in David Bohm's holographic universe theory. So let's get back to the insights that took him into the world of the hologram and the ideas that were born of that meeting. Before receiving his doctorate from the University of California at Berkeley in 1943, he worked at the Lawrence Berkeley Radiation Laboratory. It was here that he was to do his most important work on plasmas. Plasma, as you know, is a fourth state of matter and is a type of gas containing a high density of electrons and positive ions. While studying plasma, Bohm came upon a remarkable discovery. The electrons that were in plasma stopped behaving like individual entities and began behaving as though they were part of a larger connected whole. The effects they produced reflected an ability to organize themselves into what seemed like organic units that could regenerate and enclose impurities in a wall in the same way a biological organism, like an amoeba, encases a foreign substance in a cyst. He called these collective movements of electrons *plasmons*.

Universal Constants

After his work at Berkeley, Bohm went on to do research at Princeton University. There he used the knowledge he had gained with plasmas to study electrons in metals. As before, he discovered that electrons organized themselves into collective movements called **plasmons.** The significant difference between these discoveries and the Aspect experiment was that instead of just two particles knowing what the other was doing, now it seemed as though untold trillions of particles knew what all of the others were doing. It was this work that established his high regard as a physicist.

The discovery of this interconnectedness led him to question the explanation that Bohr had given about the electron's existence being dependent on it being observed. He agreed with Bohr that there was an indivisible system at work, but he refused to dismiss the interconnectedness that Bohr didn't seem to care about. He was also looking for an explanation that could account for the instantaneous communication that was being denied, but which was still an implication that needed to be addressed. Bohr and his followers claimed that quantum mechanics was complete and that it wasn't possible to arrive at a deeper understanding of what was going on at the quantum level. Bohm, of course, was strongly influenced by Einstein whom he would often spend hours with (they were both at Princeton at this time) discussing the questions that quantum mechanics didn't address.

Bohm believed that there was a deeper reality beneath the quantum level, a subquantum field he called the *quantum potential.* Based on his work with plasmas and the electrons in metals, he believed that electrons do exist in the absence of observers, the opposite of Bohr's view. And by proposing the existence of this quantum potential, he was able to explain the workings of quantum physics as well as Bohr could and also address the questions that Bohr and his followers dismissed.

The Importance of Wholeness

After Bohm developed his initial theory of quantum potential, he continued to refine it. Out of this refinement came the realization that wholeness was a significant part of the theory. In classical science, the state of a system as a whole is just the result of the interaction of its parts. But in the quantum potential, the behavior of the parts was actually organized by the whole. So this took Bohr's assertion that subatomic particles are part of an indivisible system one step further by suggesting that wholeness is in many ways the more primary reality. And it's at this point that we bring the aspect of nonlocality back into the theory. Because at the level at which the quantum potential operated, location ceased to exist. The idea of anything being separate from anything else was not part of the field, it was a wholeness.

Universal Constants

The **quantum potential** is a field that exists below the quantum level and is therefore subquantum. It pervades all of space, but unlike gravitational fields and magnetic fields, doesn't weaken with distance. Its effects are subtle, but equally powerful everywhere. Bohm published his alternative interpretation in 1952 and as expected, received a very negative response from the physics community. But he was unswerving in his convictions and even ended up publishing a book that addressed the shortcomings of science's response to new ideas.

This nonlocal aspect of the quantum potential theory also explained the connections between twin particles (as in the Aspect experiment) without violating special relativity's ban against anything traveling faster than the speed of light. Let me give you an example of the analogy that Bohm used to explain this.

Cosmonotes

Although I introduced the Aspect experiment and nonlocality previous to this discussion of Bohm's quantum potential, which contains the aspects of wholeness and interconnectedness, these experiments and the later insights gained by nonlocality came after the development of Bohm's theories. They were actually conducted to see if his theories were correct. And as the results have shown, they validate his ideas to a high degree.

Imagine an aquarium containing a fish. Imagine also that you have never seen an aquarium or a fish before (that's a big stretch, but bear with me). The only direct knowledge you have about it and what it contains comes from two television cameras, one directed at the front of the tank and the other directed at its side. As you look at the two television monitors, you might assume that the fish on each of the screens are separate entities. After all, because the cameras are set at different angles, each of the images will be different. But as you continue to watch the fish, you eventually become aware that there is a certain relationship between them.

When one turns, the other also makes a slightly different but corresponding turn; when one faces the front, the other faces toward the side. If you remain unaware of the full scope of the situation, you might even conclude that the fish must be instantaneously communicating with one another, but this is clearly not the case. There is really no communication taking place because at a deeper level of reality the two apparently different fish are one and the same. According to Bohm, the apparent faster-than-light connection between two subatomic particles is really telling us that there is a deeper level of reality we are not privy to, a more complex dimension beyond our own that is analogous to the aquarium. We view objects such as subatomic particles as separate from one another because we are seeing only a portion of their reality.

Ordered Reality

Even though this was a breakthrough in quantum theory, there was very little response to his ideas. He was at a standstill himself, not knowing where to go with his theories of wholeness and nonlocality. So he began to look at it from another perspective. Maybe it has to do more with the order of things. In classical theory, order is usually considered to be of two kinds: ordered arrangements like people, computers and crystals, and disordered things like debris left from an explosion or a box of dropped toothpicks.

Mindwarps

The example of Bohm making a connection between what appears as two different events, the ink in the glycerin and the hidden order of the universe, is called metaphoric thinking. It is a high level of abstract thinking that can relate the seemingly different aspects of events to each other through a metaphor. This process happens more often than not in problem solving, new inventions, and the discovery of new theories.

After considerable thought, Bohm realized that we are surrounded by different degrees of order, some things being much more ordered than others. This led him to the idea that perhaps order is hierarchical and that there may not be any limit to the hierarchies of order in the universe. Maybe the things we see as disordered are simply of a higher order and only appear random and chaotic. It was at this point in his thinking that he was ripe for another leap of insight. This leap occurred while watching a BBC television program. On the show was a specially designed jar that contained a large rotating cylinder. There was a narrow space between the cylinder and the jar that was filled with glycerin (a thick, clear liquid) and floating motionless in the glycerin was a drop of ink. When the handle on the jar was turned, which rotated the cylinder, the drop of ink spread out through the glycerin and seemed to disappear. But when the handle was turned in the opposite direction, the ink slowly came back together forming the original drop. This image gave Bohm the idea that order can be hidden from view (nonmanifested) as when the ink drop was spread out, or manifested, as when it reformed into the drop.

Mindwarps

The study of ordered and disordered systems led to the birth of chaos theory. This theory studies the unpredictable behavior occurring in systems that respond to deterministic laws. An essential feature of a chaotic system is that its behavior is nonlinear, meaning that a small change in initial conditions may have a very large influence on the outcome. One of the most studied systems in chaos theory is the weather. It displays all the qualities of nonlinearity, which is why meteorologists have such a hard time predicting the weather beyond just a few days.

The Holographic Universe

Shortly after this leap of insight, Bohm came across holograms and these proved to be the culminating metaphor he was looking for. In the same way that the ink drop existed in its dispersed state, the interference patterns recorded on film appeared disordered to the naked eye. Both possess orders that are hidden or enfolded from view. Well, the more he thought about this, the more he realized that the universe employed holographic principles—it was itself one giant hologram. Let's see exactly what he meant by this.

Bohm came to the conclusion that the world of our everyday lives is really kind of an illusion, like a holographic image. Underlying our reality is a deeper order of existence, a vast, primary level of reality that gives birth to all the objects and appearances in the same way that holographic film gives birth to the hologram. This deeper level of reality Bohm calls the *implicate* or enfolded order and the level at which we exist is the *explicate* or unfolded order. The manifestation of all of the forms in the universe is the result of endless enfoldings and unfoldings.

Universal Constants

Implicate and **explicate** mean respectively enfolded or hidden, and unfolded or clear. The essence of these two ideas can be found in Bohm's book, *Wholeness and the Implicate Order.* He also felt that since this process occurs endlessly between the two orders, he needed to use a different term to describe the holographic principles. Because a hologram is usually a static image, Bohm prefers to describe the universe as a "holomovement."

For example, let's take the dual nature of light. Remember that it can manifest as either a particle or a wave. According to this theory, both aspects are always enfolded in a quantum ensemble, but the way the observer interacts with the ensemble determines which aspect unfolds and which remains hidden. And if we apply this idea to nonlocality as well, we can see that when something is organized holographically, all aspects of locality break down. In the same way that each piece of holographic film contains all of the information possessed by the whole piece, this is just another way of saying that the information is distributed nonlocally. This makes viewing the universe as being made up of parts, not the real reality. In the same way that Einstein clarified that space and time are not separate but a continuum, Bohm has taken that idea and just extended it to include everything. As a holographic universe, everything is part of a continuum.

And that, my friends, is your introduction to the holomovement. But we're not done with it yet. In the next chapter we'll look at how these principles have been applied to human consciousness. I'll meet you on the holodeck!

The Least You Need to Know

➤ Fourier transforms are the mathematical equations that formed the basis for the development of the hologram.

➤ The Aspect experiment showed that subatomic particles communicate information to each other instantaneously.

➤ The quantum potential was the first attempt to define a deeper level of reality that has interconnectedness, wholeness, and nonlocality as its primary aspects.

➤ The hologram is used as a metaphor to describe the illusion of the everyday world.

➤ The holomovement is a term that describes the countless enfoldings and unfoldings that take place between the implicate and explicate orders.

Part 6

Old Endings and New Beginnings

I bet you never thought there were so many different approaches to theories about the universe. We've looked at cultural, religious, philosophic, historic, and scientific views about the origin and the structure of the universe. But there is one thing that all of these different paradigms have in common. They are all products of our human consciousness. Whether the universe was created by a divine being, originated with a big bang, has always been here, or is the result of other dimensions really has no meaning outside of the context that we put it in. So where does that leave us? Well, for one thing, regardless of the physical laws that science has discovered or the religious traditions that provide a place for us in the grand scheme of the cosmos, the universe will be here whether we are or not. After all, our planet is just a tiny little microcosm compared to the rest of the universe.

In these last four chapters, we'll focus on the aspect that all of the different paradigms we've looked at have in common, human consciousness. We'll delve into the depths of the human mind and see what we can find. We'll also take a look at symbolic systems from the East and the West to examine what they have to teach us about the universe and our relationship to it. And finally, we'll see what the religious traditions from the East and West have to say about how everything may some day come to an end.

Chapter 22

Psychology, Cosmology, and Consciousness

In the last chapter, you had a taste of how the universe can be viewed as a hologram. David Bohm's holomovement has unique aspects to it that can offer a complete, albeit unusual, description of the universe. And as I mentioned previously, in this chapter we'll look at how the metaphor of the hologram can be applied to human consciousness. And while others have contributed to these ideas, as I did in the last chapter, I plan to focus on the groundbreaking work of just one individual, Karl Pribram. His application of holographic theory to the structure of consciousness has led other researchers to gain new insights as well. But to appreciate and understand his application of holographic theory, I think it would be a good idea to provide you with a brief, basic overview of some of the main areas of psychological theory.

The Study of the Soul

While psychology developed out of the desire to understand the processes governing our mental states, it, like much of Western civilization, has its roots in ancient Greece. The root word of psychology is *psyche,* which is the Greek word for soul. And although the early questions posed by the Greeks dealt with the nature and function of the soul, those questions remained in the realm of philosophy. Psychology as we know it today is only about 150 years old. And it began with the father of modern psychology, Sigmund Freud.

Universal Constants

Psyche is the Greek word for soul; therefore, psychology translates as the study of the soul. But psychology is really more concerned with the study of the mind. The soul is more often considered a part of religion or metaphysics. The mind in psychology is considered as being subjectively perceived; that is, it's not something you can wrap your hand around. It's a functioning mechanism that is based ultimately upon physical processes with complex processes of its own. It governs our total organism and our interactions with the environment.

Freud's form of psychology is called analytical psychology, and his most important discovery was the unconscious mind. His theories are based on the belief that feelings, thoughts, and emotions that are too difficult to deal with consciously are stored in the unconscious by one of many defenses. The goal of therapy is to get in touch with the repressed material through a process called "free association" as well as through dream interpretation in order to free the person of unconscious thoughts and feelings.

Behavioral psychology was the next step in the evolution of psychology and was developed by B. F. Skinner. The fundamental belief of this system is that every human being consists of particular behavior patterns. When a particular behavior is not desired, the idea is to inhibit that behavior through a process called "operant conditioning." This process consists of removing the unwanted behavior by giving negative feedback whenever that behavior is exhibited. Eventually, through punishment and reward, the person learns to cease unwanted behavior.

The next wave of psychology, humanistic psychology, developed out of a response to the view of the human being as more than just unconscious desires and conditional behavior, as analytical and behavioral psychologies proposed. Humanistic psychology was founded by Carl Rogers and Abraham Maslow with the common theme that each person comes into the world with a desire to actualize his or her potentials as a human being. Maslow's *hierarchy of needs* has been applied to all areas of life, from organizational management to self-fulfillment. Our society is still experiencing the tremendous effects of this perspective through groups such as the Gray Panthers, National Organization of Women, etc. These groups, and thousand like them, developed because people wanted to experience their healthy and positive qualities rather than the negative qualities and pathology.

Universal Constants

Maslow believed that people needed to be understood in terms of human potential rather than just mechanical forces and unconscious instinctual impulses. He developed his **hierarchy of needs.** Here's a brief outline of the five levels, from lowest to highest:

➤ **Physiological needs**   Biological needs such as oxygen, food, water, etc., the strongest of needs.

➤ **Safety needs**   Felt by adults during emergencies, periods of disorganization; felt by children who display signs of insecurity and the need to be safe.

➤ **Love, affection, and belongingness needs**   The need to escape loneliness and alienation and to give and receive love, the need to belong.

➤ **Esteem needs**   Need for a stable, firmly based, high level of self-respect, and respect for others in order to feel satisfied, self-confident and valuable.

➤ **Self-actualization needs**   It can be described as a person's need to be that which the person was born to do, a person's "calling." Involves deep levels of devotion to something very precious and includes high levels of creativity.

The fourth wave of development in psychology, transpersonal psychology, can be considered to have evolved from the preceding three types of psychology. It incorporates many of the aspects of the three, but goes one step beyond, incorporating the idea that there is more to a person than what can be seen, felt, or experienced on a conscious or unconscious level. Carl Jung, a contemporary of Freud, broke from the traditional analytical philosophy by proposing another form of consciousness, the *collective unconscious.* Through Jung's use of symbols and his extensive explorations into the philosophies and religions of the East, he found that there is a collective world of symbols that bridge all cultures and spans lifetimes. He called these collective symbols, *archetypes,* and spent most of his life working with clients through the application of particular archetypes into their lives. Jung is considered to be one of the first applicators of transpersonal philosophy.

Cosmonotes

The four areas of psychology that are being discussed here, while investigating the workings of our minds, have also been developed as tools of "therapy." Clinical psychology focuses on helping individuals deal with their psychological problems. And each of the four modes discussed are, of course, just general definitions. There are many other theories that have been developed that are not mentioned here, such as gestalt, primal, existential, transactional analysis, and depth psychology, just to mention a few. There are also the therapies that include body alignment such as chiropractic, Rolfing, Feldenkrais, and forms of massage/energy alignment such as Swedish, Thai, Zen Shiatsu, acupressure, and Jin Shin Jyutsu that incorporate working with different psychological aspects as well. So there's a lot out there that bring together many different styles of psychological and physical therapies to promote healing and well-being.

Universal Constants

The **collective unconscious** lies at the core of our being. It is a part that we never tap consciously but only through the symbols in our dreams. It contains the highest and lowest aspects that define the best and worst of humanity. Within this "pool" of the collective unconscious are the **archetypes.** These are universal patterns or themes that everyone draws upon when they sleep. This level of the unconscious and the symbols that reside there apply to all people, cultures and races. By studying the dream symbols of people from all over the world, Jung realized that this level of consciousness is a part of everyone's psyche.

"With Our Thoughts We Make the World"

Transpersonal means going beyond just personal experience and has many times been related to mystical and spiritual experiences. Transpersonal psychology is the study and experience of expanded levels of awareness. These experiences can be mental, emotional, physical, or spiritual, but they have the characteristic of going beyond the normal, everyday experiences into realms that are expanded and expanding, to that essence of life itself. We can experience the transpersonal dimensions through bodywork and getting more in touch with physical sensations, through release in the emotional realm, through the realization of a concept never before thought or understood, or through meditation or spiritual contemplation. How we each experience reality and perceive its dimensions is a product of the space around us, the feelings we have, the thoughts we think, the beliefs we hold and the connection we have with the life force, whether that force is called God, Jesus, Atman, Chi, Tao, Mohammed, or Buddha. And in the words of the Buddha:

> We are what we think.
> All that we are arises with our thoughts,
> with our thoughts we make the world.

I thought I would throw in a little Eastern philosophy as a warm-up to what's coming up in Chapter 23, "Eastern Cosmology and the Cyclic Universe." But it's not out of context. With the idea of our minds being the interpreters of what we perceive and experience, and on a certain level create the world, we come to the theory of the mind as a hologram. At the heart of this theory, which has strong elements of the transpersonal in it, is the theory that the brain is a hologram enfolded in a holographic universe. To see exactly what this means, let's see what led Pribram to arrive at this idea.

Holographic Brains

It was thought at one time that our memory was found in localized areas of our brain. Actually some neurophysiologists still do, but there's mounting evidence to suggest that memory is distributed throughout the brain as a whole. This concept attracted Pribram to look for the mechanism that could be responsible for the distribution of memory. After reading an article about holograms, he felt he had a solution to the puzzle. Remember, one of the chief features of holographic film is that if it's cut into smaller and smaller pieces, each piece retains the image that was originally part of the whole first piece. So Pribram thought if it was possible for every portion of film to contain the whole image, it also seemed possible for every part of the brain to contain all of the information to recall a whole memory.

Besides memory, it appeared as though other brain functions operated under a holographic model, too. When large portions of the visual cortex were removed (that's the part of the brain that processes our vision), people could still perform complex visual tasks. If the brain was employing some internal holographic imaging process, even a small area of the holographic brain could reproduce the whole of what the eye was seeing.

Mindwarps

The surgery that had been performed on people's brains for medical reasons in which large sections of the brain had been removed helped to convince Pribram that something other than localized memory was responsible for memory. Patients never came out of surgery with any selective memory loss and others with head injuries never had just half a memory or forgot only portions of things. Either a complete memory was gone or the whole memory was intact.

But the central question remained. If the brain was acting like a hologram, how was it doing so? Well, it's known that the brain's structure is based on the electrical communication that takes place between neurons, the brain's nerve cells. These neurons possess branchlike endings that radiate outward and when an electrical impulse reaches the end of these branches, this impulse ripples out in a wavelike form. These constantly crisscrossing electrical waves form interference patterns very similar to the interference patterns created on a piece of holographic film.

These Are Holograms, Too?

These are some interesting adaptations of a holographic model for the brain. And just to show you that it's a plausible model, here are a few other processes that can be explained holographically:

➤ On the course of an average lifetime, our brain stores about 2.8×10^{20} bits of information. There's no mechanism to explain this vast capacity. But holograms can store incredible amounts of information. A one-inch piece of film can store the same amount of information as in 50 Bibles.

➤ Our ability to recall and forget information can be explained holographically, too. If you tilt a hologram the right way, the 3-D image appears. But if you don't turn it to the correct angle to reflect the light, the image is lost. Using this analogy, when we forget something it's because the internal holographic memory is tilted out of phase to our search.

➤ We have an uncanny ability to pick out a familiar face out of hundreds of faces. It's not based on emotion, but pattern recognition. There's a process called *recognition holography* that can explain this ability. An image is recorded in the usual way, except that in this case a laser is bounced off a focusing mirror before hitting the film. When a second object is photographed in the same way onto the same piece of film, any similarity to the first object will show up as a bright spot on the film. If there is no similarity no spot will show.

➤ Another brain function that the holographic model explains is the transference of learned skills. In other words, being able to perform a new physical movement that you've never done before. Usually it's been thought that this can be done only after repetition or when new neural connections get established. But with the interfering wave patterns of the holographic model, the brain could shift stored information around, making learning much more flexible and explain why you can move your body in a way that it's never moved before. It also explains why certain physical movements (especially in cross patterns) can actually help the brain process information more effectively.

Universal Constants

Recognition holography uses a special mirror to focus multiple images on the same film. Any similarity will show up as a bright spot on the film. If you place a light sensitive photocell behind the film, this can be used as a mechanical recognition system for security and identification purposes.

These are only some of the more common brain functions that can be explained by a holographic brain; there are many others, such as photographic memory or the phantom limb sensations that amputees feel after losing a limb.

Space and Time, in the Mind

One of the chief features that transpersonal psychology, holographic cosmology, and Eastern philosophy all have in common is that the normal everyday reality in which we live is not the most fundamental reality that we can be aware of or have access to. The discussion from the point of view of Eastern philosophy, as mentioned earlier, will be coming up in Chapter 23. But in the philosophy of the West, in the modalities of transpersonal psychology, and in the theories of the holographic paradigm, the concept of a state of consciousness or another dimension of reality is a major aspect. And although we've run across the need for other dimensions in both supersymmetry and superstring theory, these are mathematical or geometric dimensions, not ones that reflect a deeper level of reality or an altered state of consciousness. Although, as we'll see in Chapter 24, "Symbolic Systems and the Self-Aware Universe," there are correspondences that exist between these mathematical dimensions and symbolic systems like the Cabalah and the *I Ching*.

Out of the combined theories of Bohm and Pribram come two radical views to our way of thinking. These may be difficult to grasp, but I'll try to give you a good idea of how they work. These two radical views could be summarized as follows:

➤ Our brains are able to access a deeper order of existence, and in doing so, mathematically construct objective reality by interpreting frequencies that are projected from this deeper order. What this means is that our brains construct the objects we see and experience around us.

➤ This one's a little shorter but even a step further. This one states that we construct space and time. Space and time are simply states of consciousness. This, of course, has very significant ramifications for cosmology ... and I suppose for everything else, too!

Let's take these one at a time and see if we can put them into perspective. I'm going to start with the second one first. Not because it's anymore unusual sounding then the first, but because it's really not a new radical idea. It actually has its origin with a philosopher that I've made reference to before, Immanual Kant.

Categories of Space and Time

Time and space are such fundamental traits of the way in which we experience the world that it's impossible to think of experiencing anything outside of space and time. While you may be able to imagine a flower with no fragrance or a flame that's not hot, how can you imagine something outside of space and time when we have no concept of what that is like? And that is the point that Immanual Kant made when describing how we experience the world outside of us. He proposed that there are certain active organizing principles which exist in the human mind and whose function is to organize perceptions of the physical world. He called these organizing principles *categories*.

Kant said that while we think we are passively experiencing sensations from the outer world, we are really actively organizing those sensations into humanly acceptable categories. Thus when we think we are experiencing the outer world, we are really experiencing the "categories" of our own minds. A common metaphor for this description and one that we'll come across in Eastern philosophy is that the mind is simply a mirror that reflects itself. Kant felt that we could never experience the real underlying physical reality; according to him, we can only experience the categories of the human mind. And since we're not consciously aware of these categories, we don't realize that they are responsible for organizing reality in the way that we experience it.

Let me give you an example of how this could work. At this present moment you're sitting here reading this book. But what mechanism allows you to comprehend the words you're reading one by one on this page. It obviously makes sense to you, but why? In the same way that I'm writing, word-by-word, I'm not consciously aware of picking out each word to express what it is that I want to say. It just happens, and hopefully is comprehensible. Something is organizing my thoughts below the level of awareness that allows me to express an idea and at the same time allows you to comprehend the meaning.

Universal Constants

Kant's **categories** are self-organizing principles within our minds that present the experience of the world to us in a way that makes us assume that things like space, time and causality are processes that occur outside of us. We perceive cause and effect relationships because that's the way our minds are set up to interpret the sensations that it receives. The same is true of space and time. Kant argues that space and time are categories as well and that the reason we can't experience anything outside of space and time is because we have no choice but to experience the world in that way.

This then is a foundational idea of how space and time are created by Kant's categories. Instead of them being an objective, separate condition of an outer reality, in his view they are created on a level within our minds. Jung would probably place this mechanism somewhere in the collective unconscious. This is also very close to the idea expressed by Bohm's implicate and explicate order. Let's spend the remainder of this chapter bringing all these ideas together into a holographic image.

The Implicate Mind

Kant's categories are not far from Pribram's holographic brain model. And the fount of creativity and the archetypical patterns found in Jung's model of the collective unconsciousness also impact our conscious minds in ways that are beginning to be understood. This is another interpretation of an underlying dimension of consciousness that we don't have conscious access to, and this seems to be the common theme throughout these theories. And as mentioned before, this theme forms part of the core teachings of Eastern philosophy as well. Let's now take a look at the first view from a few paragraphs ago, the notion that are our brains construct the objects we see and experience around us.

Mindwarps

Our consciousness is the central mystery of our existence. No one really knows its true source or its structure. Science seeks to explain it in terms of the brain. The physical brain, with all of its chemical, electrochemical, electromagnetic, and neural networking is thought to be the source of our consciousness. The softer approaches such as philosophy and psychology seek to argue for a deeper understanding, one devoid of a strict mechanical explanation. But in any case, no one really knows what consciousness is. Whether its created by just the physical brain, or if consciousness uses the brain as the medium through which to express itself, we still have a long way to go.

For Pribram, what is "out there" is simply an immense ocean of waves and frequencies, and reality looks concrete to us only because our brains are able to take this holographic blur and convert it into the solid forms and other familiar objects that make up our world. To put it another way, it doesn't mean that basketballs and cars and everything else that we interact with are not there. It just means that all of those objects have two very different aspects of reality. When our brains filter the images through its mathematical processes, the objects manifest the way we see them. But if we could eliminate or see beneath the level at which our brains convert everything into the familiar images in space and time, we would see the frequencies that create the interference patterns that become the objects we see.

Of course, since people are part of this picture, what we've said about objects applies to us as well. So in that sense we also have two very different aspects to our reality. We can see ourselves as physical bodies moving through space or we can view ourselves as a blur of interference patterns enfolded throughout the cosmic hologram. As a matter of fact, Bohm feels that this interpretation is much closer to the way it all works. In other words, to view ourselves as a holographic mind/brain looking at a holographic universe is just another form of abstraction that attempts to break reality down into an "in here" and an "out there" separation that really doesn't exist. Remember that one of the key features of the "holomovement" is wholeness. Although it appears as though there are two separate levels of reality, there is really only one. What makes it difficult to grasp is that we're not looking at a hologram, we're part of the hologram.

Mindwarps

If you've seen the movie *The Matrix*, what we're discussing might be easier for you to imagine or understand. The idea that we are living out our lives in a world of illusion is the underlying theme of the movie. What we think is real, is simply an electronic matrix created by a complex artificial intelligence. But you never have access to the other dimension because all you know seems so real. But behind this illusion is the real reality that is possible to become aware of. You just have to take the blue pill ... and then you're not in Kansas anymore, Dorothy! (The last line will only make sense if you saw the movie.)

Past, Present, and Future

One of the most significant implications of this theory is that if space, time, and the rest of physical reality are created by the holomovement of which we are a part, then movement through space and time might be achieved by tapping into altered states of consciousness, which are also part of the enfolding and unfolding of the holomovement. We might no longer be constrained by the limitations of physical reality and have the opportunity to experience these other dimensions that exist within the holographic universe. If you wanted to visit the Himalayas, then just by thinking about them, you'd be there!

Many of the more unusual phenomena that people experience, which fall into the categories of ESP, lucid dreaming, OBEs (out-of-body experiences) NDEs (near-death experiences), and self-healing, are all naturally explained within the context of the holomovement. Even in the realm of the more "legitimate" sciences, the holographic model can offer some interesting alternatives to standard thought. Take quantum mechanics, for example. Robert G. Jahn, a professor of aerospace science and former dean of the School of Engineering and Applied Science at Princeton University has this to say:

> I think we're into the domain where the interplay of consciousness in the environment is taking place on such a primary scale that we are indeed creating reality by any reasonable definition of the term.

Mindwarps

While many of the ideas we're talking about in this section may seem pretty out there, don't be swayed by skepticism or lack of information. There is a lot of good research and reputable people investigating many aspects of what mainstream ideology considers weird or beyond respectable investigation. For example, OBEs, or out-of-body experiences, in which a person's individual conscious awareness appears to detach itself from the physical body and travel to some other location, have been reported throughout history. Jack London, Goethe, D. H. Lawrence, and Aldous Huxley all reported having this experience. They've been known and reported by the Egyptians, the tribal cultures of North America, the Hebrews, Hindus, Chinese, Moslems, Alchemists, and Greek philosophers.

Jahn believes that instead of discovering particles, physicists may actually be creating them. He supports this claim by the recent discovery of a subatomic particle called the anomalon. Its properties vary from laboratory to laboratory. This very curious event seems to suggest that the anomalon's reality depends upon who finds or creates it.

So what do you think? Are these theories too weird, too strange, or exciting and interesting? In any event, there is a lot to discover about the universe that isn't going to be explained away in a few years with a TOE, not if all factors are taken into consideration. And that thought brings this chapter to a close. In the next chapter, we'll take a look at the Eastern cosmological view.

The Least You Need to Know

➤ Psychology has developed several ways of understanding the mental processes of humanity, each type addressing different aspects of a person's psyche.

➤ The holographic paradigm can explain many of the brain's functions and can also help clarify some processes that haven't been thoroughly understood.

➤ Space, time, and causality are seen as self-organizing internal structures that create the way in which we experience the world around us.

➤ In the holographic paradigm, there is no actual separation between what's inside of us and what is outside.

Chapter 23

Eastern Cosmology and the Cyclic Universe

In This Chapter

➤ Cycles of time

➤ Cultural differences between East and West

➤ The illusionary world

➤ Correlations between Eastern cosmology and Western physics

During the 1960s there was a surge of interest in the religions and philosophies of the Eastern traditions. From the Beatles to Timothy Leary, and from philosophy and psychology to economics and politics, the ideas, teachings, and practices of Hinduism, Buddhism, Taoism, and other Eastern traditions, found their way into Western culture. Words such as karma, yoga, meditation, tai chi, kung fu, nirvana, and many others are part of our everyday language.

There has also been a blending of concepts found in physics and philosophy. With the publication of Fritjof Capra's, *The Tao of Physics*, and Gary Zukov's, *The Dancing Wu Li Masters*, back in the 1970s, parallels were drawn between quantum physics and the inherent mystical philosophy of the Eastern traditions. The perspective of these traditions, as you will see shortly, goes to the heart of cosmology and describes a system that closely resembles the chief features of the ideas being put forth here in the West regarding the necessary role our consciousness plays in any description of the universe.

What Goes Around, Comes Around

Western religion and philosophy have traditionally been separate areas of study. And even though there is some overlap, they are still distinct categories of knowledge. Western science is obviously in its own area as well, but when we study the Eastern traditions the distinctions are not as clearly marked. There is a reason for this. While the West has focused its attention on the objective, external world—science being the culmination of this method—the Eastern approach has been to explore the inner, subjective world, that of our consciousness.

When you examine the various meanings of words in a culture, you can learn an incredible amount of information that reflects the culture's knowledge base and interests. For example, there are at least fifteen different words in the Eskimo dialect for snow. Considering where Eskimos have spent much of their time, this is not surprising. The same is true of Sanskrit, the root dialect of India and other Asian tongues. There are dozens of words in Sanskrit for consciousness and its various levels of manifestation. This reflects a group of cultures that have spent over four thousand years exploring, identifying, and studying realms of consciousness that we here in the West are just beginning to understand. The point I'm trying to make is that religion, philosophy, and science in the East all come from the same source, our consciousness. And this subjective approach places humanity within the context of cosmology, rather than outside of it, as though it were just something to study.

Black Holes

Although there have been many similarities drawn between Eastern philosophy and quantum physics, parallels which we'll look at in this chapter, many physicists get very irate when "their" science is compared to such things as Buddhist or Hindu cosmology. The methods of science are very different from the understanding of Eastern perspectives, and because that approach appears so different, many refuse to accept any similarity between the two. Of course from the Eastern perspective, science is merely doing catch-up, validating ideas that have been known in the east for centuries. It'll be interesting to see what both systems of thought can learn from each other in the twenty-first century.

Besides taking a much more subjective approach to studying the universe, another defining quality of Eastern philosophy is its cyclic approach to time, space, and life.

When I say that subjectivity is the source from which knowledge comes in the East, that's not to say that they aren't also keen observers of nature. It's just that the *intention* behind the observation is different than the intention we normally have. The significant difference is the notion of wholeness. Let me explain. In most other cultures outside of, or prior to Western influence, peoples saw themselves as part of the world and ultimately the universe. The tribal cultures of North and South America, Australia, and other parts of the world understood themselves within the context of nature and its cycles. This was the observation of the Eastern traditions as well. Their intention was to understand our relationship as humans to nature and to participate as an integral part of everything, not to be separate from it.

What these observations led to was an understanding of the cyclic nature of the universe. One of the basic concepts reflected in that the yin/yang symbol mentioned in Chapter 1, "Once Upon a Time," was the fundamental cyclic nature of the universe. We live in the 24-hour day, the 7-day week, and the 365-day year, all cycles of the natural world. Every 26,000 years the earth rotates through all twelve signs of the zodiac, and beyond that our solar system cycles through the spiraling, rotating movement of our galaxy, the Milky Way.

Cosmonotes

Intention as the motivating factor to choose a particular action is different from the reason for doing something. Our reasons are often based on goals or as a way of providing a means to an end. They can be emotionally motivated, unconsciously motivated, selfishly motivated, and done out of duty or responsibility. Of course, these aren't the only reasons for doing something, but intention is a subtler form of choice. The way intention is applied in Eastern philosophy is more of a purposeful, conscious choice. It is one that is thought about for a while and doesn't seek instant gratification. In its ideal application it is a choice detached from personal investment.

Linear vs. Cyclic

This basic concept of cycles is very different from how the West views time, space, and life. We live in a linear universe. In the view of the Judeo/Christian/Islamic tradition, the universe began at a fixed point in time, the creation, and has continued in a linear fashion since then. Someday it is supposed to all come to an end as well. So the creation and the end of it are the alpha and omega, or the two end points that define the linear existence of the universe. That same theme is reflected in the most popular version of the beginning of the universe in science as well, the big bang. And as far as life is concerned, this is the pattern as well. We are born to live a linear life in time, which comes to a final end with our death. So as you can see this underlying concept of linearity is part of our culture, religious tradition, and scientific outlook. It's hard for any of us to imagine anything without a beginning and an end.

But in the Eastern traditions we find almost the exact opposite. Since the observable universe is defined by cycles that repeat themselves, this became their underlying concept for interpreting time, space, and life. It is one of the main reasons why the theory of reincarnation plays such a significant role in Eastern philosophies and religions. Why should people, who are as much a part of the cyclic rhythm of the universe, be different from everything else? Life is then considered cyclic as well. Birth and death are the cycles through which everyone and everything passes; it's called the eternal return.

Black Holes

Western religious traditions sometimes take a dim view of the Eastern concepts of reincarnation, enlightenment, and other foreign concepts. Buddhism is often mistakenly viewed as an atheistic religion. Some of the misconceptions about the Eastern traditions come from interpretations by Westerners who have a limited knowledge about the underlying meaning of certain terms, don't translate Sanskrit, Pali, or Chinese correctly, and can't comprehend the concepts being presented by these other cultures. And other material that's available is often a condensed version that doesn't do justice to the richness and complexity of these cultures. However, there is some really fine material available on these traditions, so if you're interested in finding out for yourself what any of these philosophies have to offer, talk to someone who can point you in the right direction, not someone who has already judged them as not being worth while of investigation.

Right Brain or Left Brain?

There are a couple of possible reasons why there is such a strong difference in how two cultures approach the same question. One is the way in which the cultures have developed over time and the other is where they have placed their emphasis. Linear, reason based, analytical thinking can be traced back to ancient Greece. This thinking breaks down the whole into its parts. It is the hallmark of Western culture. The other form of thinking is more synthetic and puts parts together to form wholes. It is more reflective of Eastern cultures. Interestingly enough, the two hemispheres of our brains can be shown to exhibit similar traits. Neurophysiological research has discovered that the left and right hemispheres of our brains each have their own characteristics. The following is an outline that shows how this breaks down.

Left Brain	Right Brain
Rational Drawing conclusions based on reason and facts.	**Nonrational** Not requiring a basis of reason or fact; willingness to suspend judgement.
Digital Using numbers as in counting and technology.	**Spatial** Seeing where things are in relation to other things, and how parts go together to form a whole.
Temporal Keeping track of time, sequencing one thing after another: Doing first things first, second things second, etc.	**Nontemporal** Without a sense of time.
Verbal Using words to name, describe, and define.	**Nonverbal** Awareness of things but minimal connection to words.
Analytic Figuring things out step by step and part by part.	**Synthetic** Putting things together to form wholes.
Symbolic Using a symbol to stand for something.	**Concrete** Relating to things as they are, at the present moment.
Abstract Taking out a small bit of information and using it to represent the whole thing.	**Analogic** Seeing likenesses between things; understanding metaphoric relationships.
Logical Drawing conclusions based on logic: one thing following another in logical order, as in a mathematical theorem or a well-stated argument.	**Intuitive** Making leaps of insight, often based on incomplete patterns, hunches, feelings, or visual images.
Linear Thinking in terms of linked ideas, one thought directly following another, often leading to a convergent conclusion.	**Holistic** Seeing whole things all at once; perceiving the overall patterns and structures, often leading todivergent conclusions.

All of us use both hemispheres all of the time, but individuals differ in degree of how much each side comes into play. Culturally we are much more left brain than right brain, the computer being the epitome of left-brain thinking. Eastern cultures, and again this is a generalization, are often more right brain. And their philosophy, science, and medicine reflect the more holistic approach. But as mentioned earlier, the East and the West have been interchanging cultural aspects for more than 40 years. It's common today to hear of holistic medicine and alternative therapies that embody principles found at one time only in Asia.

Unity in Diversity

To get a glimpse of Eastern cosmology we need to take a look at the ground from which it sprang. The main focus of this section will be on Hinduism. It offers a rich tapestry of philosophy of which we can examine certain threads. Although Buddhism also offers a rich heritage of insight, some of its ideas dealing with cosmology are very similar, because Buddhism grew out of aspects of Hinduism. And China and Japan as well as the other Asian countries have a lot to offer too, but we only have so much space to discuss this material. You will get a chance to take a closer look at Chinese cosmology in Chapter 24, "Symbolic Systems and the Self-Aware Universe."

Hinduism is not a single religion. That is a convenient term used in the West to describe a rather multifaceted matrix of beliefs, philosophies, practices, myths, and epics. As in the Western religious tradition, there are many stories and myths about *srishti*, the Sanskrit word for "creation."

Universal Constants

Srishti, the Sanskrit word for creation, means a projecting of a gross, unrefined thing from a subtler substance. It does not mean bringing out existence from nonexistence or creating something from nothing. The Western understanding of creation implies something arising from nothing, or nonexistence becoming existence. (Do you remember the Latin term for this concept from Chapter 17, "Scientific Origins of the Universe"? Ex nihilo ... I know you knew that!)

In Hindu philosophy, nonexistence can never be thought of as a source of creation. A more accurate description would be to say that the universe is a projection of the Supreme Being, rather than a creation. This implies that God is not separate from creation but is imminently in it. Everything that exists, all there is, is seen as a manifestation of the divine. There is nothing that does not contain a spark of the *Godhead*. All life is sacred, and our entire world and the universe are seen as a unity in diversity. In other words, since everything comes from the same source, regardless of the almost infinite number of ways it has manifested, there is an underlying principle of unity. So the most fundamental ground of being from which everything comes is essentially spiritual at the core. Here are a few sayings from Hindu philosophy that reflect that concept of unity in diversity.

There are many paths to the Top of the Mountain. Find the path that works for you, just remember that in essence they are all one and the same path.

That which is One is one in which that which is not One is also One.

Whatever divine being people choose to believe in and worship, it is only people who don't see that the source from which these beings come is all the same.

Universal Constants

The **Godhead** in Hinduism is an abstract concept that is the focal point of their sacred texts. Of course, the divine aspect of any religion is the focus of their sacred texts as well. But in Hinduism, divinity lies far beyond any human comprehension, so far that to try and describe it or talk about it, only limits what it really is. So the notion of a personal god is foreign to Hindu philosophy. However, this Godhead can manifest in various forms, which accounts for the multitude of gods and goddesses that are part of the Hindu pantheon. These are the beings that form the basis for the stories and myths of Hinduism. And even though the Godhead lies beyond comprehension, over the past four thousand years, Hinduism has developed one of the most in-depth, abstract conceptions of deity ever conceived by humanity.

The Eternal Return

The Hindu scientific approach to understanding external reality depends ultimately then on understanding the spiritual source from which everything manifests. Unlike the West, which lives in a historical world, India is rooted in a timeless universe of eternal return: Everything which happens has already done so many times before, though in different guises. This concept and all of the others that form the core of Hindu philosophy arose from the discoveries of people who felt that they had gained an insight into the nature of reality through deep meditation and ascetic practices. In the same way that science in the West has developed a methodology of examining the external world, so has Hinduism developed a methodology for examining our inner world. And for Hinduism, the real world is the one that resides inside of us. The external world is seen as illusion. So the fundamental idea underlying Hindu concepts of space and time is the notion that the external world is a product of the creative play of *maya,* illusion.

Universal Constants

Maya is the Sanskrit word for illusion. In Vedanta, one of the main schools of philosophy in India, it is the foundation of mind and matter. It manifests as two different aspects: ignorance (avidyā) and knowledge (vidyā). As ignorance or cosmic illusion, it draws a veil over our vision so that we see only the diversity of the universe rather than the unity. It leads humanity away from God and toward worldliness and imprisonment by materiality, which in turn leads to passion and greed. Its other aspect, knowledge, leads to the full realization of unity and finds expression in spiritual values. Both aspects are active in the realm of time, space, and causality.

Based on this notion of maya, the world as we know it is not solid and real, but an illusion. The universe is in a constant flux with many levels of reality. Does any of this sound familiar? These concepts are very similar to the holographic paradigm in which the world is also illusion and the wholeness and unity that exists within the realm of the implicate order is the true source from which reality unfolds. As you'll see by the end of this chapter, there are other correspondences that exist between new ideas in science and the ancient cosmology of the East.

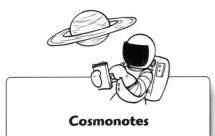

Cosmonotes

As you can see I've included the Sanskrit words for some of the concepts found in Hinduism. They're not meant to be anything more than an introduction to some of the basic concepts of Hindu philosophy. So if you can't pronounce them or get confused by their spelling don't worry. What's important is what they mean, not how they're pronounced.

A Day and Night of Brahmā

Much of Hindu cosmology is based upon the idea of endless cycles. In Hindu teachings about the universe, these cycles include a state of nonexistence before creation, the creation resulting in finite things, and a destruction ending in a return to a nonexistent state. The shortest Hindu World cycle is called the *mahā-yuga* or "great age." Two thousand mahā-yugas equals a day and night in the life of *Brahmā*. These days and nights are called *kalpas*. And there are 360 kalpas or days and nights in the life of Brahmā.

A common simile used to describe the length of time of a kalpa goes like this. Suppose that every hundred years a piece of silk is rubbed once on a solid rock one cubic mile in size; when the rock is worn away by this, one kalpa will still not have passed.

Well, that's a serious length of time. But to put all of these yugas and kalpas into a frame of reference that we're more familiar with, I'll give you an overview of how this all breaks down. Each mahā-yuga is composed of four yugas. Each one lasts shorter than the one before it in a ratio of 4:3:2:1.

➤ Krita or Satya-Yuga equals 1,728,000 human years

➤ Tretā-Yuga equals 1,296,000 years

➤ Dvāpara-Yuga equals 864,000 years

➤ Kali-Yuga equals 432,000 years

The total years for one mahā-yuga is 4,320,000 years. And one Kalpa equals 8 billion, 640 million years, making a total life span of Brahmā a little over 3 trillion years. That's a pretty old guy! And according to the Hindu calendar, we are currently in the fifty-first year of the life of Brahmā.

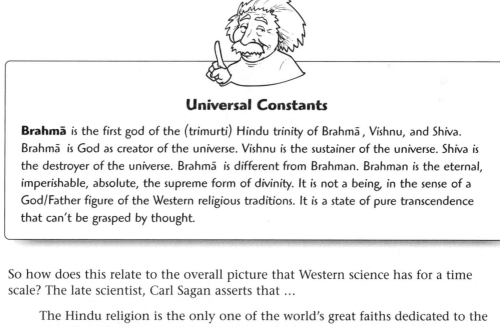

Universal Constants

Brahmā is the first god of the (trimurti) Hindu trinity of Brahmā, Vishnu, and Shiva. Brahmā is God as creator of the universe. Vishnu is the sustainer of the universe. Shiva is the destroyer of the universe. Brahmā is different from Brahman. Brahman is the eternal, imperishable, absolute, the supreme form of divinity. It is not a being, in the sense of a God/Father figure of the Western religious traditions. It is a state of pure transcendence that can't be grasped by thought.

So how does this relate to the overall picture that Western science has for a time scale? The late scientist, Carl Sagan asserts that ...

The Hindu religion is the only one of the world's great faiths dedicated to the idea that the Cosmos itself undergoes an immense, indeed an infinite number of deaths and rebirths. It is the only religion in which time scales correspond to those of modern scientific cosmology. Its cycles run from our ordinary day and night to a day and night of Brahmā, 8.64 billion years long. Longer than the age of the Earth or the Sun and about half the time since the Big Bang. And there are much longer time scales still.

In the overall scheme of the universe then, it comes into existence, is destroyed, or dies, and is reborn, over and over again. This could be the thousandth time the universe has gone through this cycle. (According to the Hindu calendar though, this is only the fifty-first time.)

The Dance of Shiva

Many of these ideas within Hindu cosmology have been represented in art. One of the most striking images ever created is the statue of Shiva. The dancing Shiva represents the constant process of creation, preservation, and destruction of the universe. The dance is a symbol of the unity and rhythm of existence. He dances in a ring of fire that refers to the life-death process of the universe. Everything is subject to continual change, as energy constantly assumes new forms in the "play" of creation, except the god himself whose dance is immutable and absolute. Images like this are found throughout India, which reflects how ancient art, philosophy, religion, and science, all come from the same source.

Mindwarps

There is an interesting correlation between Christianity and Hinduism. The first verse in the gospel of St. John reads: "In the beginning was the Word, and the Word was with God, and the Word was God." In Greek, the Word translates as Logos, which in the context of theology is the creative and sustaining spirit of God. In Hinduism, the universe comes into being from the sacred syllable **Om.** In both instances a Word is the catalyst that triggers the creation of the universe.

And it is also in this statue that we find another correspondence to Western science. Modern physics has shown us that movement and rhythm are essential properties of matter; that all matter, whether here on earth or in outer space, is involved in a continual cosmic dance. But the similarity becomes more striking when we remember that sound is a wave with certain frequencies which change when the sound does, and that particles or strings, are also waves with frequencies proportional to their energies. According to field theory, each particle does indeed "perpetually sing its song," producing rhythmic patterns of energy. So in the same way that Shiva dances, continually changing his cosmic dance to the music of the universe, matter comes in and out of existence, enfolding and unfolding for eternity.

It's All Energy

If we put the essence of Eastern cosmology into Western terms, I think you'll see why some physicists have drawn correlations between the two. On one hand we have the notion of unity in diversity. If we translate this as just energy manifesting in different forms, there is a definite correlation. The philosophical tradition of India relies on studying the spiritual source of the universe to understand cosmology, analogous to physics studying how energy originated and manifested in all its myriad forms. The Hindu concept of an eternal universe, going through its cycles of creation and destruction, is actually a combination of the big bang, plasma cosmology and steady state theory. Even spirit and matter, usually considered to be at opposite ends of physical manifestation can be understood as simply two forms of energy at different ends of the spectrum. One is just a more dense form of the other, and vice versa. The one element that binds it all together is if energy is a manifestation of some form of consciousness, or if consciousness was just another form of energy—but that, my friends, is a discussion for another time. For now, it's just some food for thought.

The Least You Need to Know

➤ Philosophy, science, and religion all have their roots in the study of the spiritual essence of the universe in Hindu cosmology.

➤ The cosmos is seen as a manifestation of unity in diversity—the veil of illusion that prevents us from seeing this is the presence of ignorance and the lack of true knowledge.

➤ In Hindu cosmology, the essential nature of the universe is based on cycles.

➤ In the West the belief in monotheism (one god) is thought to be a higher form of religion—a meaningless concept to Hinduism which sees divinity in all there is.

➤ The cosmology of Hinduism is incorporated into the stories, myths, and art of India, which symbolically represent many of the ideas of Western science.

Chapter 24

Symbolic Systems and the Self-Aware Universe

You often run across commercials on television or see ads in magazines for Tarot readings, psychic readings, astrology charts, horoscope interpretations ... the list goes on. Of course these wouldn't be advertised unless there was a public willing to pay for these services and who felt they needed answers to questions about, life, love, fortune, wealth, careers ... again the list could go on and on. Why do you think we have this need to know what the future will bring? Because that's the bottom line to all of these questions, the need to feel that what's ahead will make us happy, secure, and prepared. I'm not going to get into all of the possible psychological reasons why we seek those things, but I do want to show you in this chapter that much of what is presented today as "fortune telling" or "psychic readings" had its beginning in systems of thought that were originally intended as a way to understand the universe.

Multidimensional Symbols

Symbols and metaphors extend into the realm of everyday language and figures of speech. They also permeate images from the world of advertising as well as political slogans and emblems, the parables of our religions, the icons and writings of foreign and prehistoric cultures, legal customs and artworks, poetry and historical figures. The wedding ring, the cross, the national flag, the colors of a traffic light, the red rose, the black of mourning, the candles on a dinner table—countless objects, gestures, images, and figures of speech are linked to complex ideas and traditions. But our increasing abstractness, dependence on technology, hours spent in front of the television, in front of computers, and lives filled to the brim with not enough time to accomplish all that needs to be done seems to be drying up what was once an almost limitless flow of symbols. Of course, the language of computers as well as other aspects of our society depends on symbols for their operation, but the older intellectual systems and intuitive structures were rich in images in a way that their newer counterparts are not.

Many traditional symbols are ambiguous—they can't be explained as having a single, constant meaning. Not every dragon in every culture is an evil enemy; the heart doesn't always stand for love. In fact, the most profound symbols have different levels of interpretation and the information they provide us can give insights into the deeper levels of our psyche as well as connect us to the dynamic process of the universe. They are multidimensional, meaning that one symbol has many dimensions of interpretation and application.

A good question to ask is, "Why is symbolism so powerful?" Perhaps the main reason is that symbols communicate with the unconscious part of us. When we dream, the content of our dreams is symbolic not verbal. And as we've already discussed in Chapter 22, "Psychology, Cosmology, and Consciousness," Carl Jung's theory of the collective unconscious is the storehouse where much of the symbolic imagery resides. These archetypal symbols are very multidimensional because they contain so much information. Over the millennia and across all cultures, these archetypes have grown to include all aspects of humanity's religious, familial, scientific, cosmological, artistic, military, political, mythological, economic, philosophical, and psychological symbolism. So each archetypical pattern, and if we want to invoke the holographic paradigm, is a holographic image filled with the dynamic information of all that has ever been thought, felt, imagined, and *unknown* by all of the billions of people who have lived on this earth. And I emphasized unknown, because there is much knowledge and wisdom contained in these "implicate" patterns that we have not gained access to, at least not consciously.

Change Is the Only Constant

Millennia ago, two separate cultures, the Chinese and the Hebrew, each developed their own unique approach to synthesizing multidimensional symbols into a system. The two symbolic systems that I'm referring to are the Cabalah and the *I Ching*. We briefly touched on the Cabalah in Chapter 3, "A Brief History of Esoteric Philosophy." But we'll take a much closer, in-depth look at both of these systems in this chapter. These two systems are so old that their origin is lost in legend and myth. They're sometimes considered mystical, in the sense that they deal with aspects of cosmology, consciousness, and spirituality not found in mainstream religion or in other common areas of philosophy and psychology. They have the potential to reveal to the practitioners of these two systems a behind-the-scenes look at the knowledge and wisdom contained in the archetypical patterns we discussed earlier.

Universal Constants

The *I Ching,* or *Book of Changes,* which is the English translation of the *I Ching,* is one of the oldest known texts in the world. It supposedly dates to about 5,000 years ago. *I Ching* is pronounced *yee jing,* not "eye" ching, which is a common mispronunciation. Ching is the Chinese word for a classic book or text and the character Yee is represented by a lizard or chameleon. Hence, Ching represents the idea of change in the same way a chameleon can change its color to match its surroundings.

Energy Comes in Many Forms

About 5,000 years ago a system of signs and symbols was created in which two fundamental forms of energy, yin and yang, were combined together to form a variety of images. On a very simple level, it's a binary system, not unlike the 1s and 0s that are the basic symbols that operate all computers. As an abstract symbol, yin is represented by a broken line _ _ and yang is represented by a solid line ___. By combining these two symbols into various combinations, certain patterns were created that were assigned various qualities and attributes. The *I Ching* consists of 64 hexagrams (6-line figures) that are made up of these 2 kinds of lines. These 64 hexagrams display every possible combination of these lines when taken 6 at a time (2^6=64) But the sages who put this system together just didn't start with 6 lines. They started with just the original 2 lines and began adding lines to eventually arrive at the hexagrams. Each one described a unique pattern of energy and every time a line was added, a new energetic configuration was created. Below is an example of how it all started.

    ———   — —   ———   — —

    ———   ———   — —   — —

The first way that these two primary forces, symbolically represented by the yin and yang lines, can by combined together is illustrated above. The first combines two yang lines; the second, one of each; the third, one of each but in a reverse order; and the fourth, two yin lines, or 2^2=4. These first four combinations are considered to be the fundamental building blocks of the universe. And as we've seen before in other cultures, four again is considered a chief feature of the cosmos. Some scholars have assigned these four patterns to the four basic forces of particles physics as well. The strong force would be the first pattern, electromagnetism the second, gravity the third, and the weak force the last.

The sages then added another line to the four patterns already created, (or $2^3 = 8$) and generated the eight fundamental *trigrams* that go into making up the 64 *hexagrams*. These hexagrams are the main archetypal patterns of the *I Ching*. The following table shows the eight trigrams with some of their symbolic attributes.

Universal Constants

Trigram is a term used for the three-lined symbolic images created within the context of the I Ching. They were invented nearly 5,000 years ago by the legendary heroic figure, Fu Shi. Later these eight trigrams were combined in all possible ways to create the six-lined symbolic images called **hexagrams.** In Chinese, the word *kua* is used for both the trigrams and hexagrams. These symbolic images are the core element found in all aspects of Chinese culture.

Ch'ien	Tui	Li	Chen	Sun	K'an	Ken	K'un
⎯⎯	⎯ ⎯	⎯⎯	⎯ ⎯	⎯⎯	⎯ ⎯	⎯⎯	⎯ ⎯
⎯⎯	⎯⎯	⎯ ⎯	⎯ ⎯	⎯⎯	⎯⎯	⎯ ⎯	⎯ ⎯
⎯⎯	⎯⎯	⎯⎯	⎯⎯	⎯ ⎯	⎯ ⎯	⎯ ⎯	⎯ ⎯
the creative	the joyous	the clinging	the arousing	the gentle	the abysmal	keeping still	the receptive
the father	youngest daughter	middle daughter	eldest son	middle son	eldest son	the mother	
heaven	lake	fire	thunder	eldest daughter	water	mountain	earth
unconscious	mouth	eye	foot	wind	ear	hand	stomach
NW	pleasure	ego	movement	wood	heart	stopping	SW
	W	intellect	E	lungs	winter	NE	
		S		SE	N		

I'm not going to define the 64 hexagrams—we'd need another 10 pages just for that. But the idea behind this system is that the way in which the 8 trigrams combine to form the 64 hexagrams ends up symbolically representing all possible aspects of human and cosmic interaction. Of course, you may wonder how that's possible with only 64 figures? Each of the 6 lines has its own unique meaning in relation to the rest of the hexagram. Taken in that context and with all of the attributes assigned to the individual lines, the symbolic interpretation is almost infinite.

Mindwarps

The principles that govern the *I Ching* gave birth to its use for fortune telling. If you ask an average Chinese person about the *I Ching,* they will smile at you and say, "Oh, would you like your fortune told?" It is only the scholars and philosophers of Taoism (pronounced *dowism,* like in Dow Jones) and Confucianism who have studied its symbolism that are aware of its profound wisdom and potential for self-transformation.

The Sages of China

As mentioned earlier, the first trigrams were developed by Fu Shi. About 2,000 years later, or about 3,000 years ago, the main text and the 64 hexagrams were written by King Wen and his son, the Duke of Chou. After that, around 500 B.C.E., Confucius added his commentaries called the "Ten Wings." King Wen and Duke Chou were great sages who had a marvelous intuitive knowledge of natural principles. They used the *kuas* (the trigrams and hexagrams) as a means of helping us see into nature's ways with a view to bending ourselves to suit those ways instead of trying to conquer nature. With the *I Ching,* we learn how to adapt ourselves to fit in smoothly with nature. You see, the Taoist principle is always to swim with the current rather than against it. But even swimming with the current requires some knowledge of the current's vagaries. If you swim with an unknown current, you don't know what might happen to you. So the *I Ching* teaches us how to flow with nature and makes it easier for us to fit in.

The aim of the *I Ching* is extremely lofty, yet not at all ambitious. The ancient sages who mastered the *I Ching* did so not because they were ambitious to become great leaders and masters of men. Not at all! Their purpose was self-mastery. Why? Did they learn how to master themselves in order to become great beings? No, not at all! They learned to master themselves to be of maximum service to the community in which they lived and to the human race as a whole. King Wen and Duke Chou's part of the *I Ching* reveals how very profound the intuitions were that must have come to them as a result of tremendously high-powered meditation together with a discursive study of nature. Confucius saw in their wisdom a guide for statesmen, prime ministers, and people such as that. So his commentaries are very much concerned with how a man who has mastered the *I Ching* and mastered himself can be of use to his emperor or ruler in helping him to guide the state.

But we ourselves, as individuals, can also use the *I Ching*, not to guide the state (unless one happens to be the governor of California, for example), but because each human being is a microcosm corresponding to the macrocosm of the entire universe. Each of us inside ourselves is a symbolic representation and replica of the universe. The principles of the *I Ching* apply at the highest level to the great planets and stars wheeling in their courses and, at another level, to each individual person like you and me. That's the true meaning of a multidimensional symbol.

Mindwarps

After spending 50 years of his life studying, contemplating and writing commentaries about the *I Ching*, Confucius said at the age of 72, "If I had another 20 years to study the *I Ching*, I may begin to fathom the inner workings of the universe." A very modest statement from one of the greatest sages of China.

Cosmological Principles

Now, what are the underlying principles of the *I Ching*? The main principle is that nothing remains static. Everything in nature is subject to perpetual change. However, this change occurs in regular cycles and is governed by certain immutable laws, which however, are flexible enough to permit wide scope for us to act for better or for worse. So, if we penetrate the laws governing the movements of the universe, we can learn how things are going to happen. And knowing that, we can learn how to adapt ourselves to each forthcoming situation. In the philosophy of Taoism, the formless *Tao* produces all the forms in existence through the interplay of the archetypes or polarities of yin and yang.

Part of studying the symbolism of the *I Ching* is knowing how to interpret the words that are used to define the characteristics of the trigrams and hexagrams. For example, the heaven trigram, Chien, doesn't really mean heaven. It stands for the invisible world in which the yin and yang archetypes act together to produce the forms that are primarily represented by the so-called earth trigram, which stands for the realm of form. So "heaven" is the formless realm and "earth" is the realm we find ourselves in now. We can use the *I Ching* to trace arising situations back to their origin or forward to their completion. In this way, it's possible to learn the ways of life and to prepare for death.

Using the *I Ching* to trace these arising situations to their origin or completion is also what has given the *I Ching* the simpler notion of "fortune telling" we discussed earlier. When the *I Ching* is consulted, the energetic pattern reflected in the hexagram can reveal to you your place in the scheme of things at that exact moment in time, and project that energy pattern forward to show you the outcome. All situations result from and merge into other situations. The *I Ching* shows us how to recognize and even predict developments and thus prepare for a probable future.

Universal Constants

The closest concept we can come to in the West for **Tao,** pronounced *dow,* is the "Force" from *Star Wars.* The *Tao Te Ching,* the essential text of Taoism, contains 81 short chapters, consisting of a few short paragraphs each. It's been translated more frequently than any other work except probably the Bible.

Cosmonotes

If you're interested in finding out more about the *I Ching* there are a few good recommendations in the reading list at the back of the book. You may even find a class on it at a college near you. One of the best introductions to the study of the *I Ching* is a book called *The I Ching Workbook* by R. L. Wing. It doesn't go into a lot of the philosophy, which is a good idea when you're first learning. But it does give some history and explanation of how it works and how to consult it. Just remember to treat the whole thing with respect.

The laws that apply to the macrocosm apply equally to the microcosm. From the Taoist point of view, a human being is born of yang—in this case meaning seed, and of yin—in this case meaning power. For as long as that seed and power are in harmony, our lives go forward and we go from strength to strength. When disharmony sets in, we begin to get ill and old, and when the equilibrium is totally upset, we die. And again from the Taoist perspective, neither our origin nor our disintegration at death is absolute. It is only in a relative sense that any of us are born or die. What is real in us is unborn, undying.

The one other principle I want to briefly touch on before moving on to the Cabalah is the principle of synchronicity. Carl Jung also studied the *I Ching* for several years and discovered this principle during the course of his study. In very simple terms, synchronicity is very similar to coincidence. But it occurs as a dynamic process rather than by accident, which is a common association with coincidence. Synchronicity defines the relationship between events or situations that have nothing to do with cause and effect. For example, if you're standing in the shower singing a song to yourself, and upon getting out of the shower, turn on the radio and hear the song you were just singing, that's synchronicity. There is no direct cause and effect relationship between you singing the song and then hearing it on the radio moments later.

Mindwarps

One of the characteristics of a symbolic system is the ability it has to be related to other systems of thought and other symbols. The *I Ching,* as you've already seen, has a direct correlation to the four fundamental forces of nature. And if the 64 hexagrams are laid out in a circle in a specific order, and black and white lines are drawn, representing yin/yang, the pattern created by the swirling, spiraling lines, create images of vibrational wave patterns of subatomic particles as well as the geometric shapes of multidimensional space very similar to the forms created in string theory. It's pretty cool!

When you consult the *I Ching* for insight into a situation, the moments in which the lines of the hexagram are revealed to you, the connection you have to the universe in those moments is based on synchronicity rather than cause and effect. It's as though for those brief moments the unknown dimension of the universe, the implicate order, unfolds and the intimate part that you play in the whole, is revealed to you. It can be a moving experience and one that has the potential to teach you a lot about yourself. So what are you waiting for? The universe is waiting. Next up, the Cabalah.

The Tree of Life

Although the actual origin of the Tree of Life is unknown, it is deeply rooted in the esoteric teachings of Judaism known as the Cabalah (or Kabalah). The story is that Abraham, the father of the Hebrew nation, was given these teachings from Melchizedek, the King of Salem, a very respected priest of the time. It was then passed down through the line of priests, the Cohanim. (Many Jewish names such as Cohan, Kohen, and other spellings usually indicate that the family name is directly related to the priestly caste of Judaism, the Cohanim.)

The tree of life gives us a symbolic model of the universe much in the same way the *I Ching* has done in China. It tells us how mankind manifested out of the creative force or God into physical existence through the 22 paths or archetypes. These paths also are represented through the symbolic system of the Tarot, an ancient system that emerged from the Tree of Life and is also considered the remnants of the Book of Thoth, a teaching that escaped the burning of the library of Alexandria in Egypt. Whether these symbols came from Israel, Egypt, or China, they give us the tools for understanding our connection to the universe, to spiritual principles and a way to grow and transform as a human being.

We covered a little of the Cabalah back in Chapter 3, so if you want to refresh your memory of what the tree of life looks like, now would be a good time. The 10 spheres, or Sepheroth, have multiple correlations. As in the trigrams from the *I Ching,* each sphere can symbolically represent dimensions of the universe, male and female qualities, body parts, psychological aspects, and relate the microcosm to the macrocosm, or the individual to the whole.

The study of the Cabalah was traditionally the mystical, esoteric branch of Judaism. But as explained in Chapter 3, it has been incorporated and combined into other systems of esoteric philosophy as well. And as with the *I Ching,* I want to briefly explain some of the basic ideas and principles in the Cabalah, without overwhelming you with detail. One of the best ways to do this is to explain some aspects of the Tarot, which is in essence a Cabalistic art, and show you a little numerological manipulation, so you can have some fun with it yourself. When you work with the tarot within the context of Cabalistic teachings it's called path-working, which refers to the connecting lines or paths in the tree of life that are represented by the major arcana of the tarot.

Cosmonotes

The actual translation of the word tarot is not clear, but some of the earliest texts say that it comes from the ancient Egyptian word Ta-rosh, meaning "the royal way." Others have said that it is an anagram from the Latin word for "wheel," while still others believe that it came from the Hebrew word Torah, which is the first five books of the Old Testament, also known as the Pentateuch.

Be Your Own Psychic and Save Some Money

The tarot is divided into two main categories—the major and the minor arcana. The major arcana is where the 22 archetypal images give us the life principles or ways to progress as a human being, while the minor arcana is the four levels of experience (mental, emotional, physical, spiritual) represented by swords, cups, disks, and wands.

It is in the major arcana that our life path is revealed to us. The tarot is a symbolic system based on numerology, the study of numbers. Each one of the major arcana (22 in total) are broken down into groups of 10. In numerology, the number 10 can be broken down into 1 + 0, or 1; the number 11 can be broken down into 1 + 1, or 2, and so on. Therefore, the 22 major arcana are grouped in constellations as follows: 1, 10, 19; 2, 11, 20; 3, 12, 21; 4, 13, 22; 5, 14; 6, 15; 7, 16; 8, 17; 9, 18. Each one of these constellations holds an outer expression or personality number and an inner expression or soul number, the lower number representing the soul number and the higher number representing the personality number.

Cosmonotes

If you're interested in studying more about the psychological applications of the tarot, a great book is Angeles Arrien's *The Tarot Handbook*, listed in Appendix B, "Suggested Reading List." A psychologist and anthropologist, Angeles takes the tarot out of the Dark Ages of devils and fortune telling and gives us a picture of how it can be used as a tool for understanding ourselves.

Let's figure out what number we are in the tarot and which life path we are on this lifetime. First, get your birth date (month/day/year) and add the numbers together like you did when you were in grade school.

For example, if your birth date is 8/4/1951, you would add as follows:

$$
\begin{aligned}
&\ \ 8 \\
+&\ \ \ \ 4 \\
+&\ \underline{1951} \\
=&\ 1963
\end{aligned}
$$

Then add 1963 across, and you'll get 1 + 9 + 6 + 3 = 19. Nineteen is the first number or the personality number. Then you take 19 and add it again, 1 + 9 = 10. Hold on to this number for a minute. Then you add 10 and get 1 + 0 = 1, or the soul number. In this example, and it's the only one like this, you get a third number in the middle, the 10. This is called a gift card. And believe me, you'll need it. If your life path is the 19/10/1, your personality archetype is the Sun, the principle of teamwork, partnership and collaboration. Your soul card is the 1, the Magician and has to do with communication. Your extra card, the gift card, is 10, the Wheel of Fortune, which is the principle of turning our lives into abundance and prosperity. Your minor arcana cards would be all of the aces (1s) and the 10s, representing the different challenges and gifts that you possess. Although we can't go into detail about each one of the constellations, this gives you an idea of how it works.

Mindwarps

The playing cards we use today are said to come from the tarot's minor arcana and were used by the Gypsies as a fortune telling system during the Dark and Middle Ages. But where did the Gypsies come from? It is said that the name, Gypsy, possibly comes from "Egypt," that they were a nomadic people that originally lived in Egypt and traveled throughout Asia. When they traveled in Europe during the Dark Ages, they didn't want to reveal the more esoteric understanding of the tarot for fear of persecution, so they changed the deck into playing cards.

So, what does all this give you, some interesting conversation at your next dinner party? No, it gives you a way to understand what issues you've come in with for this lifetime and ways to work through them. It also gives you insight into why certain issues may be coming up for you at this time. Let's take our example above and move him into the future. Say your birthday is 8/4/1951, but this year (the current year as of the writing of this book is 2001) it's 8/4/2001. Each year, you move into a new archetype (still hold on to your birth archetype, though, because it forms the foundation for understanding your basic self). This next card is called your growth card and will give you the archetype that you're going through this year. It's all math, again:

$$8$$
$$+ \quad 4$$
$$+ \, \underline{2001}$$
$$= 2013: \text{Add } 2 + 0 + 1 + 3 = 6$$

Now, you're in the 6 year, or the lovers card. This is a year of working on relationships, whether personal or professional. Sometimes we get married or divorced during the lovers card since it makes us focus on what is working or not working in our relationships. You could also be working on your relationship with your inner self, wanting to know who you are and what makes you tick, so it can also be a good year to start therapy or counseling. You see, the tarot gives you a window into who you are and what issues you may be working on this year or for an entire lifetime. By understanding these issues, you can get a glimpse of your part, the microcosm, in the world, the macrocosm. So as you can see, this is also why it is often seen as nothing more then a fortune telling or is used for "psychic reading."

The Cabalah and Superstrings

Because one of the main symbolic systems contained within the Cabalah is numerology, it wouldn't be unusual to be able to relate it to modern mathematics and physics. Let's take physics and show how it relates to a subject we've already covered, superstring theory. If you remember, the fundamental building blocks of the universe are vibrating strings that can only exist in 10 or 26 dimensions. This is interesting because the *gematria* of the most sacred name of God (yod-hey-vav-hey) is 26, and Cabalistic teachings explain that the universe was created through 10 utterances (the 10 sephiroth). As you know, superstring theory postulates that there are two ways in which a string can vibrate, either clockwise or counterclockwise. Clockwise vibration utilizes 10-dimensional space and counterclockwise 26-dimensional space. Without going into some difficult Cabalistic teachings and Hebrew terminology, the ten dimensions are of course, associated with the 10 sephiroth, and the 26 dimensions break down into our four space-time dimensions and 22 hidden ones that can be related to the 22 letters of the Hebrew alphabet and the paths of the Tree of Life. Is this just a remarkable "string" of coincidences? Or is there more to this than meets the eye?

So with this introduction to the oldest symbolic systems of the East and West complete, I would like to leave you with a couple final thoughts. To what degree do these systems reflect a universe that is self-aware? Are we just creating these symbolic systems of thought like any other theory, or principle, or physical law that we assume we have discovered about the universe? What if, instead of everything being made of strings, forces, dimensions, particles, waves, and energy, everything is made of consciousness in one form or another? That idea could go a very long way in uniting science, religion, the microcosm, the macrocosm, and, of course, us into a grand unified whole! A conscious TOE! Who knows?

Universal Constants

Gematria is the ancient practice of associating Greek and Hebrew alphabets with numbers because there was no notational system for numbers like we have today. So, in the Cabalah, other philosophies, and early religions, texts could be translated into numerical equivalents. Certain names had specific numerical totals that were considered significant. It is believed by some Cabalists and other esoteric philosophers that the Bible, both the Old and the New Testament, contain a hidden numerical code that contains information about creation, spiritual principles, and other bits of knowledge and wisdom.

The Least You Need to Know

➤ A multidimensional symbol is one that is capable of unifying various systems of thought.

➤ Symbolic systems allow you to access levels of consciousness that we normally aren't able to get in touch with.

➤ The *I Ching*'s structure is based on the dynamic interplay of yin and yang and it utilizes synchronicity as an operating principle.

➤ The Cabalah and the tarot are intimately connected and one of the ways in which you can apply its principles to yourself is through path working.

➤ These symbolic systems contain the inherent idea that the universe is based on consciousness and, therefore, ultimately aware.

Chapter 25

The Future of Humanity

In This Chapter

➤ Teleology and eschatology

➤ The afterlife in ancient cultures

➤ The religious archetypes of East and West

➤ Prophecy in various cultures

➤ Biblical perspectives

We began this book by exploring the origin of the universe from the cultural standpoint of ancient peoples. So it seems fitting to end this book by covering how these same cultures believe the world will end. We've already discussed the end of the universe from the perspective of science, so it only remains to see what some religious and cultural traditions have to say.

Many traditions have prophecies about how the world will end. None of them look too good unless you have a specific belief system, in which case you'll be spared. Specific dates are often given and, in some cases, the calendars of various cultures end on a certain date as well. The major difference between all of these traditions is whether or not it's linear or cyclic. But even in the cyclic cosmologies, the end is often brought about by destruction. Of course the most familiar ideas about the "end of days" is found in the Judeo/Christian/Islamic traditions of the West. But before we look at some of those, let's see how others have seen the end of the world.

What Was It All For?

In philosophy and theology there are two areas of study that deal with questions about how humanity thinks it will all end. Teleology, from the Greek *telos* (end) and *logos* (reason), is the concept that things are directed by virtue of their ends and ultimate purposes. The other area is eschatology, a branch of theology that examines the doctrines dealing with death, resurrection, judgment, immortality, and other like topics. Let's discuss teleology first.

Cosmonotes

Frequently, teleologists have identified purpose in the universe with God's will. The teleological argument for the existence of God holds that order in the world could not be accidental and that since there is a design or plan to which all things are moving toward, there must be a designer ... God.

In cultures that have a teleological worldview, the ends of things are seen as providing the meaning for all that has happened or that occurs. If you think about history as a timeline with a beginning and an end, in a teleological view of the world and of history, the meaning and value of all historical events derives from their ends or purposes. In other words, all events are future directed. Early philosophers believed that living things have an inbuilt goal toward which they move. Aristotle's thought is manifestly teleological. He argued that all nature reflects the purposes of an immanent final cause. And, of course, because the early Church fathers were all strongly influenced by Aristotle, the Christian world-view is also fundamentally teleological; all of history is directed toward the completion of history at the end of time. When history ends, then the meaning and value of human historical experience will be fulfilled.

It's All About Us

There is a very blatant aspect about most of the beliefs concerning the end of the world. Regardless of which point of view or philosophical approach we take, whether it deals with the purpose of our existence (teleology) or actual doctrines about the last things (eschatology), it has little to do with anything outside of our world. My point being, it's not that different from the viewpoint of pre-Copernican cosmology. The significant shift that displaced humanity from the center of creation and the universe, when it was realized that the sun and not the earth was at the center of the known universe, doesn't seem to have made any difference when we discuss viewpoints about how it all will end.

We seem to have this total preoccupation with ourselves as still being the most important things in the universe. It doesn't matter which tradition we look at, except maybe the scientific, that the cycle of creation or the wrath of the God's will or whatever brings everything to an end, it happens because of us. This, of course, has to be expected, because the cultures that wrote the texts that describe their interpretation of the end of days were written when they knew very little about the rest of the universe outside their own geographical area. But has that changed any viewpoints today? It really all comes back to what you believe to be the truth. (Don't worry I'm not going to touch that one this time.) But just to show you the difference between two views, let's just briefly look at the scientific.

From the cosmological point of view of science, whether our little planet comes to an end, by whatever means, really will have no impact on a universe that is infinite in size. As we've previously discussed, any scenario that science has to offer, be it the big bang, the steady state, or the plasma universe, we have little to do with the galactic picture. Even long after our sun has burned out, the universe will continue on for hundreds of billions of years, if not for eternity. So when religions or cultures discuss the end of the world, or the end of time, it's important to remember that they're referring to human civilization and human time, not the earth, the animals, or other living things—just us.

Cosmonotes

The study of eschatology is a branch of theology that deals with Western religious traditions. The word derives its meaning from the Greek *eschatos*, the last things. And while Judaism and Islam have their version of the last days, it is in Christianity where the term was first used to refer to the doctrines established by the Church that deal with the many events that relate to the final days of the world.

Ethical Eschatology

To get back to humanity and how we see our time coming to an end, there is a fundamental aspect to all points of view that is common. How we live our lives decides our fate. In other words, our ethical conduct has a direct impact on what happens to us after we die. Death and what happens after death is the focal point of all religious traditions, East and West. Divine justice and moral retribution, however, are only aspects of the Western traditions. The Law of Karma is the great equalizer in the Eastern traditions. And again we also deal with the view of linearity or cycles. For in the Western traditions, there is only the one life while in the Eastern traditions there are many lives. But for both traditions, your ethical conduct determines what happens to you. In the West if you lived a virtuous life, heaven is your reward. In the East, a good life generates a future life that has its own rewards. But before we take a closer look at eschatology in the West, let's look at how some earlier cultures viewed the end of life.

The Underworld

The ancient cultures of Mesopotamia, which include Babylonian, Assyrian, Scythian, and other Middle Eastern cultures, had some characteristics common to the civilizations that developed centuries later. The main feature was the belief in an afterlife. And if you lived a good life, you were rewarded with strength, prosperity, long life, and numerous offspring. However, if you were wicked, you would be punished by calamities in your life. It dealt mostly with the present life rather than rewards or punishment in the afterlife.

But the existence in a hereafter was believed in. A kind of semi-material ghost, or shade, or double, survived the death of the body. After the body was buried or cremated, the ghost would descend to the underworld to join the company of the departed. This underworld is described in gloomy colors and in phrases like, "the land of no return," "the house of darkness," and the "place where dust is their bread, and their food is mud." Not too cheerful sounding. It also appeared that there was no penalties for the wicked or reward for the good. Good and bad were involved in the same dismal fate.

Mindwarps

The belief in life after death seems to be one of the oldest beliefs of any culture. Anthropologists have found skeletons of Neanderthal people roughly 50,000 to 80,000 years old that revealed that they used to bury their dead with garlands of flowers and stone tools. The bodies were laid in a fetal position, which could indicate that they were being born into a new life. Great care was always taken in the burial and the idea of burying the dead with implements from life became a practice of many other cultures.

Mummies, Mummies, and More Mummies

There was probably no other culture in the history of the world that was as obsessed about the afterlife as the Egyptians. Huge portions of the society were involved in temple building, tomb carving, embalming, while others worked to support these workers. The greatest monuments on earth were devoted to the protection, worship, and continuing existence of the deceased. The ultimate goal after death was unending life with Osiris, who journeyed through the underworld on a daily basis. The departed were habitually called the "living," their sarcophagus, "the chest of the living." And it is not merely the disembodied spirit that continues to live, but the soul with certain bodily organs and functions suited to the conditions of the new life.

It is one of the reasons why so much time and energy was spent preserving the body through mummification. This art reached such a high level of success that some mummies over 5,000 years old still retain human features, skin, and hair. The reason why the art of mummification reached such a high level of skill was directly related to the beliefs about the afterlife. So long as the physical body remained intact, the Ka, or spirit, remained alive on the other side. The Ka could return to the physical body and rest in it at times. Burial chambers would contain food and drink along with many items that were used while alive. The spirits never reached a state in which they were not dependent on earthly aids.

Mindwarps

The conception of the underworld was a common theme in many ancient cultures. It was found in Mesopotamia, Egypt, and in Greece. Each civilization had their own myths and stories associated with how one traveled to get there, what it was like, and who inhabited those realms. It wasn't like any modern notion of heaven or hell. Just some region "under the world."

Universal Constants

The Egyptian **Book of the Dead** is a misnomer applied by historians to a text that the ancient Egyptians referred to as the Book of Coming Forth by Day. No definitive version of this book exists. Rather it is a compilation of funerary texts and religious hymns written by priests and copied by scribes during a period spanning approximately 4500 B.C.E. to 200 C.E. The Egyptians were a much more spiritually evolved civilization than early archeologists thought. And recent interpretations of papyrus show them to be much more technologically advanced as well.

Each soul had to undergo a judgment that was described in detail in the *Book of the Dead*. The examination covers a great variety of personal, social, and religious duties and observances. The deceased must be able to deny his guilt in regard to 42 great categories of sins, and his heart (the symbol of conscience and morality) must stand the test of being weighed in the balance against the image of Maat, the goddess of truth and justice.

Thus Spoke Zarathushtra

The one other ancient culture we're going to look at is Persia. Some of the prominent features of both Judaism and Christianity can be found here prior to their inception. The main source of the teachings comes from a prophet called Zarathushtra, or Zoroaster, as he is called in the West. Many scholars think he lived sometime between 1300 and 600 B.C.E. Although there are few living representatives of Zoroastrianism (few compared to most other religions), for over a thousand years it was the official religion of the vast Persian empire that stretched from Turkey to India. Zoroaster introduced the following beliefs that were later integrated into Jewish, Christian, and Muslim religions:

➤ The belief of heaven and hell

➤ An evil force or being

➤ Judgment of the individual and resurrection of the body after death

➤ A dramatic apocalyptic end of the world with a final resurrection of the dead

The Sage and the Prophet

With the syncretization of the most basic eschatological elements of Zoroastrianism into the Judeo/Christian/Islamic traditions, prophecy would also be incorporated as a significant instrument to further explain the outcome at the end of time. And this prophetic tradition also describes a significant difference between Eastern and Western approaches to life, death, salvation, and the afterlife.

If we were to compare the basic archetypes of the Eastern and Western religious traditions, we would see two unique patterns forming which reflect each tradition's essential focus. The archetype of the Eastern religions is the sage. The sage is seen as the wise person who embodies wisdom, compassion, understanding, intuition, and self-efforting as the way to achieve salvation. Only salvation is again very different in the East compared to the West. In most of these Eastern traditions, *transcendence* is the goal. Salvation is not achieved by belief in a spiritual being, although that is an aspect of some faiths, but by working on and becoming aware of the restrictions, limitations, attitudes, behaviors, beliefs, and actions that prevent us from achieving spiritual understanding and self-realization. All of these aspects need to be ultimately transcended before one can realize and recognize the innate spiritual essence that resides at the core of our being.

Universal Constants

It's often difficult for adherents of the Western religions to understand exactly what **transcendence** is about. What in the world needs to be transcended and how will that bring salvation? In 25 words or less it's impossible to relate the goal of the Eastern traditions, so I'll just make one or two statements. What needs to be transcended is the ego. It goes by different names in various traditions: the false self, the lower mind, the player on the stage, the monkey mind, the veil, and that which experiences separation, to name a few. For in these traditions, the ego is the source of all problems, for it needs to think that it is in control and real. Union with God (Hinduism), the extinguishing of the ego (Buddhism), and at-one-ment with nature (Taoism) all see the ego as the thing in the way of attaining the spiritual potential each seeks to offer.

In the Western traditions the archetype is the prophet. Although Christianity considers Jesus as the one and only Son of God, he in many ways became the ultimate prophet, not speaking in place of the Divine, but speaking as the Divine. However, from both the viewpoints of Judaism and Islam, he is still considered a prophet. And of course in Judaism, the prophets play a key role, and in Islam, Mohammed is the "Seal of the Prophets."

The prophet was a person who spoke forth for God, not so much as a foreteller but as a forthteller, not one who told future events, but one who spoke in the place of the Divine ("Thus Says the Lord"). When someone today is referred to as a prophet or is said to prophesy, we think of a soothsayer, someone who foretells the future. This was not the original meaning of the word. "Prophet" comes from the Greek word *prophētēs* in which *pro* means "for" and *phētēs* means "to speak." Thus in the original Greek, a prophet is someone who "speaks for" someone else. And a prophet differed from other men in that his mind, his speech, and occasionally even his body could become a conduit through which God addressed immediate historical conditions.

Mindwarps

The prophetic movement in Judaism passed through three stages, the first one being the stage of the Prophetic Guilds, of which the ninth and tenth chapters of First Samuel provide one of the best glimpses. In this stage prophecy was a group phenomenon. Prophets are not here identified as individuals because their talent is not an individual possession. Traveling in bands or schools, prophecy for them was a field phenomenon that required a critical mass. Contemporary psychology would consider it a form of collective, self-induced ecstasy. With the help of music and dancing, a prophetic band would work itself into a state of frenzy. Its members would lose their self-consciousness in a collective sea of divine intoxication.

A Prophecy Primer

With the prophets firmly established in the Western traditions, we now turn to prophecy itself. Prophecy not only plays an important part in the Judeo/Christian/ Islamic tradition, but is a phenomenon that can be found in other cultures as well. There are also some famous people who have gone down in history, known for their prophetic gifts. One of the most well known, and one that has become very popular over the past 25 years is Michel de Nostra-Dame, otherwise known as Nostradamus. There have been numerous movies made about his life and predictions, and his quatrains have found their way into many books about his life and the future he saw for humanity.

Prophecy plays a role in other cultures, too, most notably in the Mayan, Hopi and other ancient and pre-modern tribal societies. In the Mayan culture, prophecy was directly linked to the Mayan calendar. In some respects, Mayan cosmology is similar to the Hindu cycles in that they cover immense periods of time. But the calendar that we're going to look at covers one Great Cycle from 3113 B.C.E. to 2012 C.E. It's broken into 13 divisions, each covering approximately 385 years. We are currently in the last period, with only a short time left before the calendar ends. The entire system is very similar to the *I Ching,* too, since it's based on the interaction of different types of symbolic energy. It remarkably predicted the rise and fall of Mayan culture to within one year. The final period we're in now is described as being one of tremendous change and upheaval, followed by centuries of peace and global prosperity.

Mindwarps

In all, Nostradamus wrote 1,000 verses that involve 2,500 predictions. Of these about 800 have supposedly already been fulfilled, and others relate either to the future or to events which cannot be checked. Nostradamus came from a Judaic family lineage, but had been baptized Catholic. He wrote most of his predictions during the middle of the 1500s in cryptic verse so that he wouldn't get into a lot of trouble with the Church. Some people put a lot of credence to his work, while others remain skeptical. It's always easier to interpret the meaning of prophecy after it has happened then to actually sight an exact event or date in the future. This is true of all prophecy.

The Hopi also have prophesies about this time in human history, but it's not based on a calendar. Rather, it is linked to the visions that many Hopi Medicine teachers had. And again it, too, speaks of this period in history of earth changes, troubled economic and political problems. In most of the tribal cultures there was always a shaman, medicine person, or holy man who would act as the spiritual guide of the tribe. In order for them to know what was best for their people, they would enter into altered states of consciousness that would enable them to experience the realm of the "spirits" to obtain visions of what to do. These visions would often be prophetic in nature, warning of events to come and offering guidance and wisdom to the people.

Cosmonotes

One of the most enduring and prophetic tribal holy men was Black Elk of the Ogalala Sioux. The biography of his life and his incredible prophetic visions are all clearly presented in the book *Black Elk Speaks* by John G. Neihardt. His insight into the plight of the tribal cultures of North America is very profound. And his prophecies of Custer's last stand as well as the events in the twentieth century are incredible. It's a great read!

The Christian Outlook

From the Christian point of view, prophecy is often considered proof that the Bible is the literal word of God. Since the prophecies that have been fulfilled are considered to be accurate to the finest detail and since no human writer could be that accurate, the conclusion is reached that they all must be divinely inspired. There were over 300 Old Testament prophecies about the Messiah that are believed to have been fulfilled by the life of Jesus. Prophecies that have not yet been fulfilled are often the ones of greatest interest to Christians. But the thing that makes the study of prophecy challenging is that there are so many different trends and ways that events can be interpreted.

Not all Bible scholars agree on how future prophecy will be fulfilled. There are some who think that future prophecies will not be fulfilled literally, but that they merely have spiritual meaning. Others, who do expect a literal fulfillment of all the things to come, sometimes disagree on the timing of certain events. Jesus' own teaching about the future is found in his Sermon on the Mount of Olives. He indicated that there would be four phases of the future. Here's an outline of the four phases, taken from various sources in the New Testament and interpreted as literal history:

➤ **The signs of the times** This is considered to be the times that we are currently living through. It began with the foundation of the nation of Israel in 1948 and includes events such as: earthquakes, famines and plagues, wars and rumors of war, lawlessness, persecution, terrors in the heavens and false Christs.

➤ **The tribulation** During this time, the Antichrist will come to power, first over the revived Roman Empire, and then over most of the planet. This will be made possible supposedly by the growing interest by earth's citizens in a one-world government or new world order. (This is a Christian interpretation of events and simply reflects the beliefs and the desire to "fit" historical events into that particular frame of reference.)

➤ **The great tribulation** Jesus describes this time as one of "great distress." The Antichrist defiles the Temple by setting up an image of himself and demanding the worship of all. He will utilize modern technology to control an economic dictatorship so that only those who worship him and receive the "Mark of the Beast" (666) will be allowed to buy or sell. The last three years of the Tribulation are covered by the Seven Trumpets and the Seven Vials of God's wrath. (This is covered in the book of Revelation.)

➤ **After the tribulation** Jesus returns to earth in the Second Coming. He will destroy the forces of evil at the Battle of Armageddon. After that there will be a thousand years of peace, followed by one last rebellion by Satan. Finally after the last battle, God will create a "New Heaven and Earth."

That's a general breakdown of how some Christians see the "end of days." On an individual level, the belief in Jesus Christ is the most important thing to have. Without it, one will not gain entrance to the "Father's House." The belief in the resurrection of Jesus guarantees individual resurrection after death, but individual judgment still comes into play. So how one lives, regardless of what one believes, is still of paramount importance. There are many other aspects to Christian Eschatology. And since there are over 50,000 denominations of Christianity worldwide, that also reflects a wide variety of perspectives of beliefs and attitudes to how everything will end. But regardless of your personal beliefs there does seem to be a point of commonality among all the various traditions. Christian, Mayan, Hindu, Hopi, Sioux, Jewish, Muslim, and a few others all believe that within the first 20 years of the new millennium, there are going to be some tremendous upheavals and change; some of it good, some of it not so good. And how each of us deals with whatever the future holds depends on how much we can express what is best in humanity rather than what is worst. So with that I would like to close our look at the theories of the universe by a poem by William Blake:

> To see a World in a Grain of Sand
> And a Heaven in a Wild Flower,
> Hold Infinity in the palm of your hand
> and Eternity in an hour.

The Least You Need to Know

➤ Teleology and eschatology are the two areas of study that examine ideas and beliefs about the future of humanity.

➤ Salvation is the goal of the Western religious traditions and transcendence the goal of the Eastern religious traditions.

➤ The sage represents the religious archetype of the East and the prophet the archetype of the West.

➤ Prophecy is an aspect of many traditions that is used to validate, forewarn, and prepare the adherents of these traditions for what the future holds.

Cosmological Kernels

acupressure   Similar to acupuncture, this ancient Chinese art is based on the treatment of disease and pain through finger pressure on points on the body along energy channels.

acupuncture   An ancient Chinese healing art that uses needles to stimulate points along energy lines (called organ meridians) to relieve pain and treat disease.

agnostic   A person who believes that the human mind cannot know whether there is a God or ultimate cause beyond the material.

alchemy   From the Greek word "to pour," it was an early form of chemistry practiced in the Middle Ages that attempted to change baser metals into gold, literally and spiritually as well.

alpha decay   During radioactive decay, it is the process that happens when the nucleus of an atom ejects an alpha particle.

amber   A yellow or brownish-yellow translucent fossil resin.

amplitude   In a wave, it is half the height from the peak of the wave to the bottom.

analogy   The process of describing similarity in some respects between things that otherwise are unlike.

analytical geometry   The branch of geometry in which position is indicated by algebraic symbols and solutions are obtained by algebraic analysis.

angstrom   A unit used in measuring light waves, it is one hundred millionth of a centimeter in length.

angular momentum   A term used to describe the momentum of a rotating object around another object in a curved trajectory.

anthropomorphic Conceiving or representing a god with human attributes or assigning human qualities to nonhuman things such as animals.

antimatter A type of matter that has the opposite property to its counterpart in our world (i.e., an antielectron or positron has the opposite charge, positive, to an electron which has a negative charge).

archetype The term used by Carl Jung to describe the symbols and mythic figures that reveal inherent psychological processes of a universal nature.

Aristotle A Greek philosopher (384–322 B.C.E.) and student of Plato who is most noted for his works on logic, metaphysics, ethics, politics, etc. He established the idea that the material world is made up of four elements—air, earth, fire, and water.

astrology The study of the stars and planets and their influence on humanity.

astrophysics The science of the physical properties and phenomena of the stars, planets, and all other heavenly bodies.

atomic time The standard time used today based on atomic clock data and starting from 0 hours, minutes, and seconds GMT on 1/1/58.

baryon A term used to describe a fermion that follows the strong force (i.e., proton and neutron).

beta decay The process by which a neutron within the nucleus of an atom, or a free neutron, ejects an electron.

black body radiation The radiation emitted by a black body, a hypothetical object that can absorb all electromagnetic radiation that it comes in contact with.

black hole An area of space where matter becomes concentrated with enough gravitational field to curve space-time upon itself so that nothing can escape, not even light.

Book of the Dead From Egyptian antiquity (written between 4500 B.C.E. and 200 C.E.), it is the book that spelled out the funereal rituals, procedures, and hymns to be used following death.

boson The category of particles associated with the transmission of forces (the photon carries the electromagnetic force).

Brahmā The first god of the Hindu trinity, who is seen as the creator the universe.

carbon-14 dating The process of dating plants and animals based on the absorption of carbon-14, a radioactive form of carbon, which is produced by the interaction of cosmic rays with nitrogen in the atmosphere.

cartesian coordinate system A pair of numbers that locate a point by its distances from two intersecting, often perpendicular, lines in the same plane.

categories The term used by Immanual Kant to describe how we experience the world through organizing principles in the mind.

cathode   A negatively charged electrode in a battery or vacuum that repels electrons due to their negative charge.

centrifugal force   The force tending to pull an object outward when it is rotating rapidly around a center.

Chaldea   An ancient province of Babylonia in the region of the lower courses of the Tigris and Euphrates rivers.

chi   The Chinese term used for describing the invisible energy that moves through the body.

compactification   The term used in the Kaluza-Klein theory to describe the spatially compressed location of the fifth and higher dimensions.

complementarity   In quantum physics, it is the term given to certain pairs of variables that prevent both of them from having precise values at the same time, such as a wave and a particle.

Confucianism   The Chinese doctrine combining philosophical, religious, and sociopolitical aspects reflected in the teachings of Confucius, in which the state reflects a high moral and social order.

conservation of electrical charge   The electrical charge that occurs when a neutron decays to form a proton and an electron.

Copenhagen interpretation   From the 1930s to the 1980s, it was the explanation of how the quantum world worked, primarily from the efforts of Niels Bohr who worked in Copenhagen.

cosmic rays   High-energy particles from space that collide with the nuclei of atoms in the atmosphere of the earth and produce showers of secondary cosmic rays.

cosmogony   The origin or generation of the universe.

cosmography   The science dealing with the structure of the universe as a whole and its related parts—geology, geography, and astronomy are branches of cosmography.

cosmology   The branch of philosophy and science that deals with the study of the universe, its form, nature, etc. as a physical system.

creationism   The doctrine that God created the universe and everything in it as described in the Book of Genesis.

cuneiform   The first written form in which wedge-shaped characters were used. It was developed in ancient Babylonia.

cyclotron   An accelerator which sends particles in a spiral direction in order to kick particles to a higher energy. The first cyclotron was built by Ernest Lawrence at Berkeley in 1930.

deductive   The process of reasoning from a general to a specific conclusion.

dogma   A doctrine or body of doctrines formally and authoritatively affirmed, sometimes without proof.

Druid   A member of a Celtic religious order of priests, soothsayers, judges, poets, etc. in ancient Britain, Ireland and France.

electromagnetism   Discovered in the nineteenth century by James Clerk Maxwell, it is the process of unifying the forces of electricity with magnetism.

electron   The negatively charged particle that forms a part of the atom.

element   In chemistry, it is a substance that cannot be separated into different substances by ordinary chemical methods.

energy   It is the product released when an object does "work," such as the heat released when exercising.

epistemology   The study or theory of the origin, nature, methods and limits of knowledge.

eschatology   A branch of theology that deals with the actual doctrines of the end of the world.

esoteric   A body of knowledge that was known by a chosen few and kept hidden from the general populace.

ether   An imaginary substance that was regarded by early scientists as filling all space between the heavenly bodies.

exoteric   A body of knowledge that is reflected in the beliefs of the general populace.

explicate   Used by David Bohm to describe the world as we see it rather than the hidden, or implicate, world.

fermion   The category of quantum particles that make up the material world (electron, proton).

Fourier transform   The mathematical equations that convert images into waveforms and back into patterns or images (like your television).

frequency   A term used for the number of cycles per second that electricity oscillates.

frequency threshold   The name given to the way a piece of material reacts to light by emitting a certain number of electrons. The lower the electrons emitted, the lower the frequency threshold.

gematria   It is the ancient practice in Greek and Hebrew alphabets of ascribing numbers to letters.

general theory of relativity   An expanded version of Einstein's special relativity theory which explains why all bodies, regardless of their mass, fall freely in a gravitational field with the same acceleration.

gluon   The carriers of the force that holds quarks together.

Gnosticism   A system of belief that is derived from Greek philosophy, Eastern religions, and esoteric Christianity that stresses intuitive knowledge (gnosis).

Godhead   In Hinduism, it is the abstract meaning of the creative force that can manifest in many different forms (Brahmā, Vishnu, Shiva being a few of them).

gravitational mass   The measure of the amount of matter in a body determined by its gravitational force.

graviton   The term given to the theoretical particle which carries the force of gravity.

hadron   A particle that responds to the strong force and comes from the Greek word for strong.

heterotic   From the Greek for "two," it refers to the two different dimensions ascribed to fermions (10 dimensions) and bosons (26 dimensions).

hexagrams   The six-lined symbolic images of the *I Ching*.

hierarchical theory of needs   Developed by Carl Rogers and Abraham Maslow, it is the level of needs ascribed to human awareness and development.

Higgs field   The term given to the force field that fills the Universe which has a definite magnitude (strength) but no direction.

hologram   The image that is projected in three dimensions from a two-dimensional picture projected through laser beams.

hyperbola   One of three common curved figures found in analytical geometry.

I Ching   Translated into English as the *Book of Changes,* it is the symbolic system used in China for understanding our place in the world.

implicate   Used by David Bohm to describe the deeper order of existence that is hidden or enfolded that gives birth to our visible or explicate reality.

inductive   A type of reasoning that develops a conclusion based on facts.

inertia   The tendency of matter to remain at rest or, if moving, to stay in the same direction.

isotope   A similar form of an element but without the same atomic weight.

lepton   The name given to the six elementary particles that are not quarks and do not feel the strong force. All leptons are fermions.

linear accelerator   A particle accelerator which accelerates in a straight line.

M-theory   The term coined by John Schwarz to describe the unification of the various string theories under one theory. Rather than acting like strings, all of the particles are rolled up in dimensions of membranes.

major arcana   In the symbolic system of the tarot, this represents the 22 archetypal roles that human beings can express.

maya   The Sanskrit word for "illusion"; it is what keeps us from seeing ourselves as part of everything, including God.

meson   The group of particles (quark and antiquark) that are bound together by the exchange of gluons and are members of the boson family.

microcosm   A term used to describe the small, microscopic world in contrast to the macrocosm, the larger universe of which we are a part.

modular functions   Developed by the Indian mathematical genius Srinivasa Ramanujan, it is the term used to describe his brilliant equations that provide the theoretical basis for string theory.

muon   A member of the lepton family, the heavy counterpart to the electron.

nanosecond   One billionth of a second.

neutrino   A lepton with a 0 charge and a spin of ½. Neutrinos respond to the weak interaction.

neutron   One of the elementary particles with a 0 electric charge and a spin of ½.

node   The end points of a vibrating wave.

nonlocality   The instantaneous communication between subatomic particles.

optics   Relating to the eye or sense of sight.

paradigm   A pattern, example, or model.

particle physics   The area of physics that studies the tiniest matter, particles.

perennial philosophy   A phrase coined by Gottfried von Leibnitz which refers to the ancient underlying philosophy that forms the basis for all the religions of the world.

pion   A particle that belongs to the strong force (i.e. the force between protons and neutrons).

Planck's constant   The fundamental constant that relates the energy of a quantum of electromagnetic radiation to its frequency.

plasma   The fourth form of matter (the others being solid, liquid and gas) in which electrons have been stripped from atoms to leave positively charged ions.

plasmons   A term used by David Bohm to describe the movement of electrons in plasma as they form organic units that work together to regenerate and protect one another.

proton   An elementary particle found in the nucleus of all atoms that carries a positive charge. The atomic number of an atom is equal to the number of protons in its nucleus.

psyche   The Greek term for the soul.

quanta   The term given to the smallest particles in the world.

quantum foam   The term coined by John Wheeler to describe what happens when the different forces of nature interact under ultra-microscopic magnification.

quantum leap   The discontinuous transition between quantum states that occurs when an electron leaps from one energy level into another.

quantum mechanics   The laws of mechanics that apply to the quantum world.

quantum potential   A field that is smaller than the quantum level, which pervades all of space but doesn't weaken like gravitational and magnetic fields.

quark   An elementary particle that forms the basic building block of all hadrons. It's a level of matter below protons and neutrons.

redshift   The effect seen in a spectrometer, created by a wavelength of light, that shows that the object being measured is moving away from you.

refraction   An optical illusion that occurs when light rays are bent as they pass through water or any medium of a different density.

relativistic mass   When in motion, it describes the mass of an object (or particle) measured by an observer who is moving relative to the object as opposed to stationary or rest mass.

resonance   The ability of one vibrating body to set in motion or amplify another body.

rest mass   The mass of an object (or particle) that is measured by an observer who is not moving relative to the mass. The rest mass of a particle traveling at the speed of light would have 0 rest mass even though it is always in motion.

second law of thermodynamics   This law states that heat is always lost when energy moves from a higher to lower state.

singularity   Refers to the single point, infinitely small and infinitely dense, that was the beginning point of the big bang.

space-time continuum   In relativity theory, it is the union of three-dimensional space with the fourth dimension, time.

sparticle   In supersymmetry theory, it is the name given to the particle counterparts in everyday matter (electron becomes selectron).

special relativity   The theory developed by Einstein that describes frames of reference that are moving in a straight line, toward or away from each other, are not accelerating, but instead are moving at a constant or uniform speed.

spin   In quantum mechanics, it is the term used to describe the rotation of quanta, similar to the Earth rotating on its axis (although quanta don't always rotate in this fashion).

srishti   The Sanskrit word for "creation."

string theory   Any type of theory that describes the subatomic particles and their interactions in terms of very small, one-dimensional strings.

strong force   The force within the nucleus of an atom that holds the nucleus together.

supersymmetry   The theory developed in the 1970s that brought together the two different patterns of behavior in nature (particles and forces) in one geometrical framework (SUSY for short).

Tao   The Chinese word that describes the dual nature of reality into forces called yin and yang.

teleology   From the Greek *telos* (end) and *logos* (reason), it is the concept that there is a reason or meaning for the end of things.

transcendence   To go beyond something to another level of understanding or existence.

transpersonal   The fourth wave of psychology that involves a human being looking beyond just the physical, mental, and psychological processes and incorporate the spiritual.

trigram   The three-lined symbolic image from the *I Ching* that forms the foundation for the hexagram images.

uncertainty principle   Developed by Werner Heisenberg, it is the discovery that you can't pinpoint the position of a subatomic particle while knowing its momentum and conversely, you can't determine the particle's momentum while knowing its position.

velocity   The rate and change of position in relation to time (i.e., the speed of a car tells you how fast it is going, the velocity tells you in which direction).

wave/particle duality   In the quantum world, it describes how a quantum entity can act like a wave or like a particle depending on what you are looking for.

wavelength   The distance from peak to peak in a wave (or trough to trough).

Suggested Reading List

Alchemy, Hermeticism, and Esoteric Philosophy

Fauvel, John, editor, Raymond Flood, contributor, and Robin J. Wilson, editor. *Let Newton Be! A New Perspective on His Life and Works*. Oxford, England: Oxford University Press, 1989.

Hauck, Dennis William. *The Emerald Tablet: Alchemy for Personal Transformation*. New York: Penguin Putnam, Inc., 1999.

Holmyard, E.J. *Alchemy*. Mineola, New York: Dover Publications, Inc., 1990.

James, Jamie. *The Music of the Spheres: Music, Science and the Natural Order of the Universe*. New York: Springer-Verlag, 1993.

Jurriaanse, Aart. *Bridges: Basic Studies in Esoteric Philosophy*. Silverton, South Africa: Sun Centre School of Esoteric Philosophy, 1985.

Merkel, Ingrid, and Allen G. Debus, editors. *Hermeticism and the Renaissance: Intellectual History and the Occult in Early Modern Europe*. Cranbury, New Jersey: Associated University Presses, 1988.

Principe, Lawrence M. *The Aspiring Adept: Robert Boyle and His Alchemical Quest*. Princeton, New Jersey: Princeton University Press, 1998.

Shumaker, Wayne. *The Occult Sciences in the Renaissance: A Study in Intellectual Patterns*. Berkeley: University of California Press, 1973.

Teeter-Dobbs, Betty Jo. *The Janus Faces of Genius: The Role of Alchemy in Newton's Thought*. Cambridge, England: Cambridge University Press, 1991.

Yates, Frances A. *The Rosicrucian Enlightenment*. Boulder, Colorado: Shambhala Publications, Inc., 1972.

———. *Giordano Bruno and the Hermetic Tradition.* Chicago: University of Chicago Press, 1964.

Consciousness and Brain/Mind Research

Edelman, Gerald M., and Giulio Tononi. *A Universe of Consciousness: How Matter Becomes Imagination.* New York: Basic Books, 2000.

Goswami, Ph. D., Amit. *The Self-Aware Universe: How Consciousness Creates the Material World.* New York: Jeremy P. Tarcher, Putnam Books, 1993.

Hobson, J. Allan. *Consciousness.* New York: W. H. Freeman and Company, 2000.

Lynch, Aaron. *Thought Contagion: How Belief Spreads Through Society.* New York: Basic Books, 1996.

Mindell, Ph.D., Arnold. *Quantum Mind: The Edge Between Physics and Psychology.* Portland, Oregon: Lao Tse Press, 2000.

Wilbur, Ken, Editor. *Quantum Questions: Mystical Writing of the World's Great Physicists.* Boston: New Science Library, 1985.

Zohar, Danah. *The Quantum Self: Human Nature and Consciousness Defined by the New Physics.* New York: William Morrow and Company, Inc., 1990.

Cosmology and Quantum Physics

Greene, Brian. *The Elegant Universe: Superstrings, Hidden Dimensions, and the Quest for the Ultimate Theory.* New York: Vintage Books, 1999.

Guth, Alan H. *The Inflationary Universe: The Quest for a New Theory of Cosmic Origins.* Cambridge, Massachusetts: Perseus Books, 1997.

Hetherington, Norriss S., editor. *Cosmology: Historical, Literary, Philosophical, Religious, and Scientific Perspectives.* New York: Garland Publishing, Inc., 1993.

Kaku, Michio. *Hyperspace: A Scientific Odyssey Through Parallel Universes, Time Warps, and the 10th Dimension.* Oxford: Oxford University Press, 1994.

Kane, Gordon. *Supersymmetry: Unveiling the Ultimate Laws of Nature.* Cambridge, Massachusetts: Perseus Publishing, 2000.

Kragh, Helge. *Cosmology and Controversy: The Historical Development of Two Theories of the Universe.* Princeton, New Jersey: Princeton University Press, 1996.

———. *Quantum Generations: A History of Physics in the Twentieth Century.* Princeton, New Jersey: Princeton University Press, 1999.

Lerner, Eric J. *The Big Bang Never Happened: A Startling Refutation of the Dominant Theory of the Origin of the Universe.* New York: Vintage Books, 1992.

Livio, Mario. *The Accelerating Universe: Infinite Expansion, the Cosmological Constant, and the Beauty of the Cosmos.* New York: John Wiley & Sons, Inc., 2000.

Mitchell, William C. *The Cult of the Big Bang: Was There a Big Bang?* Carson City, Nevada: Cosmic Sense Books, 1995.

Rees, Martin. *Before the Beginning: Our Universe and Others.* Cambridge, Massachusetts: Helix Books, 1997.

———. *Just Six Numbers: The Deep Forces That Shape the Universe.* New York: Basic Books, 2000.

Weinberg, Steven. *The First Three Minutes: A Modern View of the Origin of the Universe.* New York: Basic Books, 1993.

Holograms and the New Physics

Bohm, David. *Wholeness and the Implicate Order.* London: Routledge, 1980.

Capra, Fritjof. *The Tao of Physics.* Boston: New Science Library, 1985.

Nadeau, Robert and Menas Kafatos. *The Non-Local Universe: The New Physics and Matters of the Mind.* Oxford: Oxford University Press, 1999.

Talbot, Michael. *The Holographic Universe.* New York: Harper Collins Publishers, 1991.

Zukav, Gary. *The Dancing Wu Li Masters: An Overview of the New Physics.* New York: William Morrow and Company, Inc., 1979.

Symbolic Systems

Arrien, Angeles. *The Tarot Handbook: Practical Applications of Ancient Visual Symbols.* Sonoma, California: Arcus Publishing Company, 1987.

Baynes, Cary F. *The I Ching or Book of Changes.* The Richard Wilhelm Translation rendered into English. Princeton, New Jersey: Princeton University Press, 1967.

Case, Paul Foster. *The Tarot: A Key to the Wisdom of the Ages.* Richmond, Virginia: Macoy Publishing Company, 1947.

Fortune, Dion. *The Mystical Qabalah.* New York: Red Wheel/Weiser, 2nd Ed., 2000.

Gray, William G. *The Ladder of Lights.* York Beach, Maine: Samuel Weiser, Inc., 1968.

———. *The Talking Tree.* New York: Noble Offset Printers, Inc., 1977.

Halevi, Z'ev ben Shimon. *Kabbalah: Tradition of Hidden Knowledge.* London: Thames and Hudson, 1979.

———. *Tree of Life: An Introduction to the Cabalah.* London: Rider & Company, 1972.

Hoeller, Stephan A. *The Royal Road: A Manual of Kabalistic Meditations on the Tarot.* Wheaton, Illinois: The Theosophical Publishing House, 1975.

Moran, Elizabeth, and Master Joseph Yu. *The Complete Idiot's Guide to the I Ching.* Indianapolis: Alpha Books, 2002.

Ni, Hua Ching. *The Book of Changes and the Unchanging Truth.* Los Angeles: College of Tao and Traditional Chinese Healing, 1983.

Wing, R.L. *The Illustrated I Ching.* New York: Dolphin Books, Doubleday & Company, Inc., 1982.

Index

353

T

X–Y

Z